NANOSTRUCTURED MATERIALS - FABRICATION TO APPLICATIONS

Edited by **Mohindar Singh Seehra**

Nanostructured Materials - Fabrication to Applications
http://dx.doi.org/10.5772/65164
Edited by Mohindar Singh Seehra

Contributors

Celal Kurşun, Musa Gogebakan, M. Samadi Khoshkhoo, Jürgen Eckert, Abouelmaaty Aly, Maria Ines Tavares, Paulo Sergio Rangel Cruz Silva, Emerson O Silva, Livia R Menezes, Jose Alberto Alvarado, Zachary Garbe Neale, Hector Juarez Santiesteban, Antonio Arce-Plaza, Juan Pablo Morán-Lázaro, Florentino López-Urías, Emilio Muñoz-Sandoval, Oscar Blanco-Alonso, Marciano Sanchez-Tizapa, Alejandra Carreon-Alvarez, Héctor Guillén-Bonilla, Maria De La Luz Olvera Amador, Alex Guillén-Bonilla, Verónica María Rodríguez-Betancourtt, Alexey Vereschaka, Sergey Grigoriev, Nikolay Sitnikov, Gaik Oganian, Raul Valenzuela, Romain Brietweiser, Ulises Acevedo, Souad Ammar, Mohamad Hafiz Mamat, Ahmad Syakirin Ismail, Mohamad Rusop Mahmood, Ratna Naik, Maheshika Palihawadana-Arachchige, Vaman Naik, Prem Vaishnava, Bhanu P Jena, Tomasz Goslinski, Dariusz Mlynarczyk, Tomasz Kocki, Ramalingam Chidambaram, Evy Alice Abigail

Published by InTech
Janeza Trdine 9, 51000 Rijeka, Croatia

Publishing Process Manager Dajana Pemac
Technical Editor SPi Global
Cover InTech Design team

First published July, 2017
Printed in Croatia
Legal deposit, Croatia: National and University Library in Zagreb

Additional hard copies can be obtained from orders@intechopen.com

Nanostructured Materials - Fabrication to Applications, Edited by Mohindar Singh Seehra
p. cm.
Print ISBN 978-953-51-3371-1
Online ISBN 978-953-51-3372-8

For EU product safety concerns:
IN TECH d.o.o., Prolaz Marije Krucifikse Kozulić 3, 51000 Rijeka, Croatia,
info@intechopen.com or visit our website at intechopen.com.

Meet the editor

Professor Mohindar Singh Seehra received his early education in India (BSc degree in Punjab University and MSc degree in Aligarh University). He joined West Virginia University as an assistant professor of Physics in 1969 after receiving his PhD degree (University of Rochester, NY) and became an associate professor in 1973, a professor in 1977, and an Eberly distinguished professor in 1992. His honors in research include A. P. Sloan Foundation Research Fellow (1974–1976); Fellow of the American Physical Society (1984–present); Fellow of the Institute of Physics, UK (2001–present); and Outstanding Referee Award of the American Physical Society in 2010. He has guided the research of 65 graduate and postdoctoral students and authored over 300 publications on the magnetic, dielectric, optical, and catalytic properties of materials and nanoparticles. In 2016, he became an Eberly professor emeritus.

Contents

Preface

A nanostructured material has at least one of its dimensions in the 1 nm to 100 nm range (1 nm = 10^{-9} meters). So, thin films with thickness less than 100 nm, wires with diameter D less than 100 nm, and particles with diameter D less than 100 nm are examples of nanomaterials (NM). A nanocomposite is a composite of an NM with any other material. During the past 20 years or so, interest in the properties and applications of NM has increased enormously among researchers worldwide in diverse fields of physics, chemistry, engineering, and biomedicine. This interest stems from very interesting and changing properties of NM with changes in the size of the nano-dimensions and potential applications of these properties to many diverse areas such as catalysis, energy conversion processes, sensors, magnetic storage media, and biomedicine.

The interesting size-dependent properties of NM result from a combination of finite size effects, role of the different properties of the surface atoms vis-à-vis those of the atoms in the core of NM, and quantum confinement effects in really small NM. Compared to atoms in the interior, atoms on the surface of a material have comparatively weaker bonding with their neighbors because of the reduction in the number of the nearest neighbors at the surface, leading to different properties of the surface atoms. For a bulk material, the fraction of atoms on the surface is quite negligible, and so their effect is not observed in the measured properties. However, for an NM, the fraction of the surface atoms to the total number of atoms in the NM increases inversely with nano-dimension D, and so the role of the surface atoms becomes quite dominant. As examples, for a spherical nanoparticle of D = 100 nm, the percentage of atoms on its surface is about 1%. However, this percentage increases to about 25% for D = 5 nm and 40% for D = 3 nm. The finite size effects result from the fact that NM with dimensions less than the characteristic length scale associated with a particular phenomenon has different size-dependent properties. Examples of size-dependent properties are electronic structures, energy band gaps, melting temperatures, electrical and thermal conductivity, and magnetic ordering temperatures. For magnetic nanoparticles, below a critical size, nanoparticles become superparamagnets with entirely different static and dynamic magnetic properties, which in turn lead to many applications. In semiconductors, quantum confinement effects and discrete energy levels appear as the size becomes comparable to the exciton Bohr radius, leading to the formations of quantum wells, quantum wires, and quantum dot analogous to thin films, nanowires, and nanoparticles, respectively.

In recent years, a number of books and review papers have been published in these emerging areas of nanoscience and nanotechnology, which readers can access for background and earlier developments. Some examples of the published books in this category are *Introduction to Nanotechnology* by C. P. Poole and F. J. Owens, *Magnetic Nanoparticles* edited by N. T. K. Thanh, and *Handbook of Nanostructured Materials and Nanotechnology* (five volumes) edited by H. S. Nalwa. Since new results continue to emerge in this fast-moving research area, there is an ongoing need to have accessible research and review papers that researchers worldwide can access. This is the primary justification of this new book on nanostructured materials. The eleven chapters contained in this book are written by different research groups from around the world who are willing and able to share their research and expertise in their particular research area in an open access format.

The eleven chapters are organized in two sections: Section 1 contains four chapters dealing with the fabrication of different nanostructures followed by their structural characterization and Section 2 contains seven chapters focusing more on the applications of nanostructures along with their fabrication and characterization. The four chapters in Section 1 are arranged as follows: Chapter 1 deals with different methods of preparing ZnO nanostructures with different morphologies and dopants; Chapter 2 is on the fabrication of ferrite nanostructures by the very new technique of spark plasma sintering, which allows to produce full densification of ferrite particles in the nm range; Chapter 3 is on the phase and size evaluations of mechanically alloyed Cu-Mg-Ni powders with the formation of a CuMgNi ternary alloys with FCC structure and grain size less than 20 nm; and Chapter 4 describes results from the investigations of the usefulness of nanoscale multilayered composite coatings deposited on carbide cutting tools, enhancing their useful life by several folds. In these chapters, the focus is on the fabrication of the NM by different techniques followed by structural characterization using techniques such as X-ray diffraction, scanning electron microscopy, and transmission electron microscopy.

For the seven chapters in Section 2, there is further hidden subgrouping since three chapters on biomedical applications and two on sensors are grouped together with the remaining two chapters on solar cells and herbicides placed at the end. Among the three chapters on biomedical applications, Chapter 5 provides an excellent review on superparamagnetic iron oxide nanoparticles (SPION) including methods for their fabrication and structural, morphological, and magnetic characterization relevant for their use in theranostics (a single platform that combines therapy and diagnosis of cancer). The applications of SPION for drug delivery, magnetic resonance imaging, and magnetic hyperthermia therapy (MHT) are included along with recent results on MHT study on Gd-doped SPION as a possible theranostic agent. The chapter concludes with a discussion on the challenges and outlook on the future research for theranostics in clinical settings. Chapter 6 is a review of dendrimer nanostructures that are spherical nanosized polymers with a core from which branched polymeric dendrons diverge and they are end-capped with selected terminal groups. This chapter reviews selected modifications of dendrimers for their use for the prevention and fighting off infections caused by bacteria and fungi, viruses, and parasites/protozoa.

Chapter 7 is a review of polymer nanocomposites based on polyvinyl alcohol (PVA), polycarbonate (PC), and matrixes of dental resins along with nanoparticles of TiO2, silver, and carbon nanotubes. Characterization of the nanocomposites using microscopy, X-ray diffraction, thermal analysis, and low-field NMR relaxometry is discussed along with their applications in dentistry and drug delivery systems.

Chapters 8 and 9 deal with the fabrication, characterization, and applications of NM systems in sensor technology. In Chapter 8, the authors review the synthesis of $ZnCo_2O_4$ nanostructures by microwave-assisted colloidal method followed by characterization by X-ray diffraction, Raman spectroscopy and microscopic techniques and its application as sensor for the detection of CO gas. The authors of Chapter 9 describe the synthesis and characterization of aluminum and iron-doped ZnO nanoarrays and their applications for humidity sensing. The Al-doped ZnO showed dense arrays with smaller nanorod diameter and highly porous surface yielding higher electrical conductivity and more than two times higher sensitivity as a humidity sensor compared to undoped ZnO nanorod arrays.

Chapter 10 deals with a review of the progress that has been made to date to enhance the efficiencies of various nanostructures used in solar cells including utilizations of all the wavelengths present in the solar spectrum. The performance evaluation of various solar cells including those based on quantum wells, quantum wires, quantum dots, and multi-junctions is covered. Finally, the authors of Chapter 11 describe the intriguing use of nanoparticles as carriers for overcoming the problem of herbicide resistance of weeds in agriculture. The use of polymeric, inorganic (silicon dioxide), and rice husk nanoparticles as carriers for herbicides to overcome the weed resistance is discussed.

It is indeed amazing and gratifying to see all these new methods of fabrication of nanostructured materials and their potential applications in many diverse areas of human interest. As editor of this book, I thank these authors for their contributions and for their willingness to make changes to their manuscripts, sometimes major, after my editorial recommendations and suggestions. However, credit for the quality of the research presented in these chapters really belongs to these authors. Also, a uniform style of writing was not possible under this process, and so readers may find somewhat varying style of writing from chapter to chapter. An important feature of these reviews and research articles is the list of up-to-date references that readers can access in these specific areas. It is hoped that new as well as seasoned researchers will find these chapters of great use in their research.

Mohindar Singh Seehra
Department of Physics and Astronomy
West Virginia University
Morgantown,
West Virginia, USA

Section 1

Fabrication and Characterization

Chapter 1

ZnO Nanostructures Synthesized by Chemical Solutions

Jose Alberto Alvarado Garcia, Zachary Garbe Neale,
Antonio Arce-Plaza, Avelino Cortes Santiago and
Hector Juarez Santiesteban

Additional information is available at the end of the chapter

http://dx.doi.org/10.5772/intechopen.68278

Abstract

Nanomaterials have been synthesized using several different techniques. Some of these techniques are sophisticated, expensive and need certain training before use. However, there are other highly efficient methods for preparing nanomaterials that are easy to work with and require no specialized equipment, making them relatively inexpensive routes for synthesis. The least expensive routes are those that are classified as solution-based techniques such as colloidal, sol-gel and microwave-assisted synthesis. The focus of this chapter is on a general description of each technique with recent advances in synthesis, doping processes and applications. Specifically, these processes are discussed in connection with the synthesis of ZnO compounds and its related nanomaterials.

Keywords: ZnO, synthesis, chemical solutions, nanostructures

1. Introduction

An important II–VI semiconductor is ZnO which has been well-studied and applied in a variety of applications. It has a band gap of 3.6 eV and large exciton binding energy of 60 meV. Nowadays this material is considered as one of the most important large band gap semiconductors due to its easy synthesis, stability at room temperature, eco-friendly properties, being a direct band gap material and fast mobility. This material exists in three different crystal phases such as zinc blende, cubic or rock salt and wurtzite or hexagonal. The first two phases are obtained only in certain well-controlled conditions such as certain pressures and on specific substrates. However, the most common phase under ambient conditions is the wurtzite hexagonal crystal structure shown in **Figure 1**.

Another advantage of this compound is that it can be synthesized and deposited by employing different techniques. Slight variation in process conditions can result in different product morphologies and properties. Since the costs associated with research and industry is always an important consideration, it becomes necessary to use inexpensive and efficient methods to obtain the desired novel nanostructured materials with applications in different fields such as optoelectronics, solar cells, piezoelectric and sometimes in biological materials.

Sol-gel, colloidal solution and microwave-assisted synthesis are techniques that are still important in the synthesis of semiconductor nanomaterials. These techniques share some similar characteristics such as (i) they are relatively inexpensive; (ii) the efficiency of the synthesized materials is high; (iii) process parameters are easily controlled and (iv) these techniques are also well-studied. For these reasons, in this chapter we have focused on a review of these techniques, especially for the synthesis of ZnO, with emphasis on the recent advances in the synthesis of novel nanomaterials and its applications. A general overview of each process is also presented for ease of readability. The synthesized materials have been structurally characterized using X-ray diffraction (XRD) and scanning electron microscopy (SEM). **Figure 2** shows a representative XRD pattern of ZnO. XRD patterns of synthesized material can be compared to reference patterns to determine phase purity or if there is preferential crystal orientation. Most of the time, ZnO is obtained as a polycrystalline film or powder which can be identified by its numerous diffraction peaks at relative intensities. Depending on the processing conditions, single crystal or preferential growth can occur

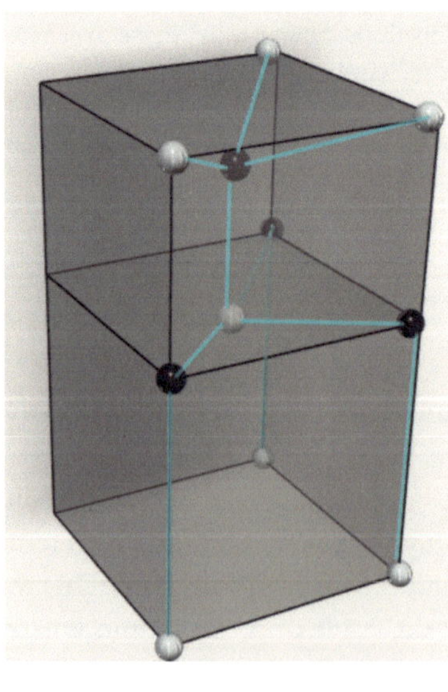

Figure 1. Representation of ZnO wurtzite crystal structure (black and grey balls corresponds to Zn and Oxygen atoms).

in thin films that result in different relative peak intensities or missing peaks compared to the reference pattern. The 2-theta values of the (100), (002) and (101) lines in **Figure 2** of the hexagonal crystal planes are located at 31.770, 34.422 and 36.253° for wurtzite ZnO (Ref. JCPDS card # 36-1451).

Different processing parameters may result in different microscopic product morphologies of ZnO. From SEM, we can observe that this material could be obtained as nanoparticles

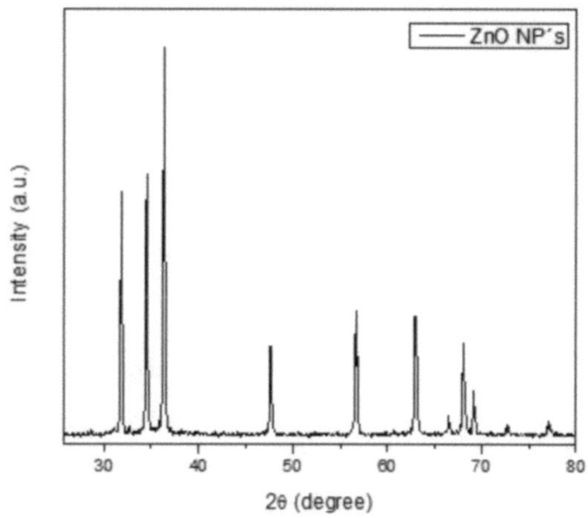

Figure 2. Typical XRD pattern of ZnO nanoparticles.

Figure 3. SEM image of ZnO nanoparticles obtained via colloidal synthesis. The scale bar is 500 nm.

(**Figure 3**), polycrystalline (**Figure 4**) and as a nanostructured thin film (**Figure 5**). All of these materials were synthesized under non-extreme conditions using colloidal synthesis to produce the source material. The crystal structure of these materials is the hexagonal wurtzite structure.

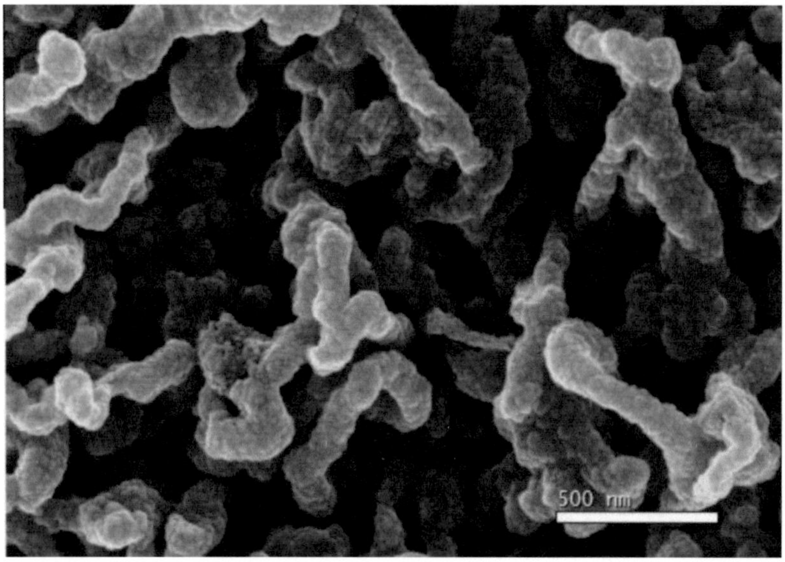

Figure 4. SEM image of polycrystalline ZnO thin film obtained through vacuum evaporation process, colloidal nanoparticles as source were used. The scale bar is 500 nm.

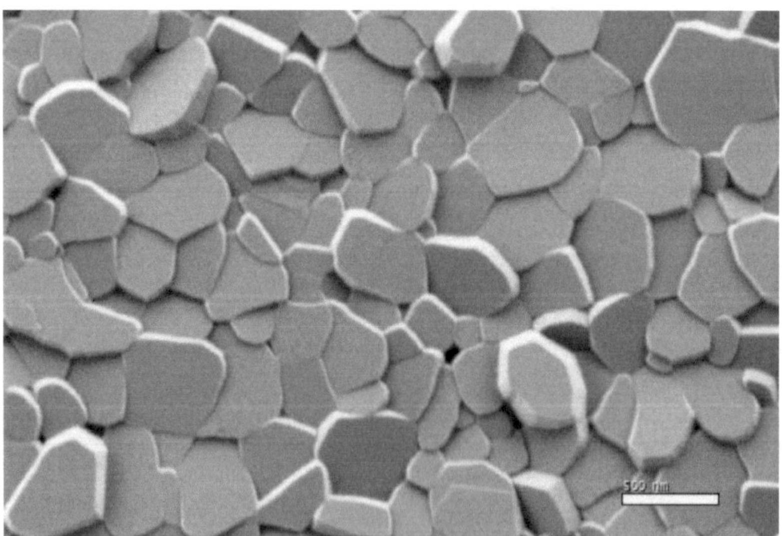

Figure 5. SEM ZnO nanostructures using colloidal nanoparticles as source. The scale bar is 500 nm.

2. Some techniques for synthesizing ZnO nanostructures and nanoparticles

2.1. Sol-gel

The sol-gel process encompasses a variety of precursors, solvents and additives. But in general, the basis of the sol-gel process includes some form of hydrolysis and condensation reactions. In the case of ZnO, usually a zinc salt such as zinc acetate is used with water or an alcohol as the solvent. An example of possible hydrolysis and condensation reactions for ZnO are shown in Eqs. (1) and (2), where $Zn(OR)_2$ is a soluble salt.

$$Zn\,(OR\,)_2 \underset{\rightarrow}{H_2O} Zn\,(OH\,)_2 + 2R \tag{1}$$

$$Zn\,(OH\,)_2 + Zn\,(OH\,)_2 \rightarrow (OH\,)Zn - O - Zn(OH\,) + H_2O \tag{2}$$

During the hydrolysis reaction, the soluble zinc precursor forms a zinc hydroxide intermediate that is able to condense with other intermediates to grow a zinc oxide inorganic polymer. The final product after drying has an amorphous structure and crystallization of ZnO particles require an annealing step. The morphology of the inorganic network can range from spherical nanoparticles to percolated gels and is highly dependent on the choice of precursors, water content, solute and solvent ratio, aging and additives. The sol-gel process has proven to be an inexpensive and relatively simple method of ZnO nanoparticle synthesis that is tailorable to produce unique nanostructures for different applications.

2.2. Colloidal solution

Colloidal synthesis is another well-known chemical solution method to obtain novel nanomaterials with different morphologies and sizes. All processing conditions involved in the system can be fixed to control nucleation and growth of the materials. The kind of interactions (physical and chemical) between particles include Vander Waals, electrostatic, Ostwald ripening and some other theoretical principles such as Derjaguin, Landau, Venvey and Overbeek theory (DLVO). These interactions can contribute to agglomeration and subsequently precipitation of the particles. Colloidal instability can be prevented through steric stabilization which usually requires a surfactant to maintain the colloidal suspension. Surfactants work in two ways: first, to prevent particulate interactions and second, to prevent the continuous nucleation and growth of particles.

2.3. Microwave-assisted synthesis

Microwave-assisted synthesis is a relatively recent technique that has been used for synthesis of nanomaterials. It has been considered as a promising approach to obtain novel nanomaterials in organic and inorganic fields. Additionally, microwave synthesis is considered as a green process and coheres perfectly to the principles formulated by Anastas et al. related to green chemistry [1].

Often a domestic microwave is used and the synthesis is carried out in solvent-free solutions. This technique allows for rapid and homogeneous heating of the system since energy is transmitted directly through molecular vibrations. The short heating ramp time of microwave synthesis allows for better control of particle size distribution compared to conventional heating. On the contrary, the extremely high heating rate of microwave-assisted synthesis may cause the boiling point of the solution to increase by a few degree Celsius. Additionally, the microwave susceptibility will vary between different materials and temperatures.

The microwave energy is generated by a magnetron that transforms electrical energy into a strong magnetic field. The electromagnetic energy interacts with the solution, vibrating the molecules and giving sufficient activation energy to the system for chemical reactions to take place in seconds or minutes.

The reaction rate during microwave synthesis can be explained through the Arrhenius equation [Eq. (3)] as follows:

$$K = A\, e^{-\Delta G/RT} \tag{3}$$

where K is the rate constant, T is the absolute temperature (in Kelvin), A is the pre-exponential factor, a constant for each chemical reaction that defines the rate due to frequency of collisions in the correct orientation, ΔG is the activation energy for the reaction (in Joules) and R is the universal gas constant. Thus, the two parameters affecting the kinetics of a particular chemical reaction are temperature and activation energy.

Bilecka et al. reported that nanoparticle growth can be described using four thermodynamic parameters related to the Arrhenius equation through activation energy [2]. These variables are the activation energies for precursor solvation, monomer formation, nucleation and crystal growth. As with colloidal synthesis, nucleation and growth in microwave synthesis are governed by Ostwald ripening.

3. Synthesis of ZnO nanostructures and nanoparticles via chemical solutions: recent advances

Sol-gel, colloidal and microwave-assisted synthesis are effective techniques to efficiently obtain novel ZnO nanostructures. These techniques are relatively inexpensive and do not require sophisticated laboratory equipment. Additionally, slight variations in precursors or process parameters can produce different morphologies that can be applied in different technological fields.

3.1. Process, materials and precursors

The precursors used in these synthesis routes usually start with a basic salt of Zn, a solvent and a catalyser such as temperature. The Zn precursor must be soluble in the selected solvent such that it can provide the necessary Zn ions to produce ZnO particles. Other reagents may be added in order to substitutionally dope ZnO with metal cations such as Fe, Cu, Co and Ba. Additionally, surfactants may be added to maintain colloidal stability of the product or influence the morphology of the growing particles.

Different precursors used in sol-gel and colloidal techniques from recent publications have been summarized in **Tables 1** and **2**, respectively. The readers are asked to consult the relevant publications for details of these processes.

Precursor	Solvent	Stabilizing agent	Reference	Technique
$Zn(CH_3OO)_2$ $2H_2O$	CH_3OH, C_2H_5OH, C_3H_7OH, C_3H_7OH, C_4H_9OH	$(CH_2CH_2OH)_2NH$, $N(CH_2CH_2OH)_3$	Pourshaban et al. [3]	Sol-gel
$Zn(CH_3COO)_2$ $2H_2O/CuCl$	2-methoxyethanol	$(CH_2(OH)·CH_2·NH_2)$	Joshi et al. [4]	Sol-gel
$Zn(CH_3OO)_2$ $2H_2O$, $Ba(NO_3)_2$	2-methoxyethanol	$(CH_2CH_2OH)_2NH$/DEA	Kasar et al. [5]	Sol-gel
$Zn(CH_3OO)$ $2H_2O$, $(NH_4)_2CO_3$, $Fe(NO_3)_3$	Distilled water/ethylene glycol	–	Bahari et al. [7]	Sol-gel
$Zn(CH_3OO)_2$ $2H_2O$, $Mn(CH_3CO_2)_2$ $4H_2O$	Isopropyl alcohol	Urea	Kumar et al. [6]	Sol-gel
$Zn(CH_3OO)_2$ $2H_2O$, $C_2H_3LiO_2$	C_2H_5OH	$(CH_2(OH)·CH_2·NH_2)$	Boudjouan et al. [8]	Sol-gel
$Zn(CH_3OO)_2$ $2H_2O$, $CaCl_2$	CH_3OH, C_2H_5OH	–	Slama et al. [9]	Sol-gel
$Zn(CH_3OO)_2$ $2H_2O$, $(CH_3COO)_2·Co$ $4H_2O$	CH_3OH	Mono ethanolamine $(CH_2(OH)·CH_2·NH_2)$	Dhruvash et al. [10]	Sol-gel
Zn(CH3COO)$_2$ $2H_2O$	C_2H_5OH	–	Singh et al. [21]	Sol-gel
Zn(CH3COO)$_2$ $2H_2O/KOH$	CH_3OH	–	Kim et al. [22]	Sol-gel
Zn(CH3COO)$_2$·$2H_2O$	2-methoxyethanol	$(CH_2(OH)·CH_2·NH_2)$	Tabassum et al. [11]	Sol-gel
Zn(CH3COO)$_2$·$2H_2O$/Al(NO3)3 $9H_2O/AgNO_3$	C_2H_5OH	Diethanolamine (DEA)	Khan et al [12]	Sol-gel
$Zn(CH_3OO)_2$ $2H_2O$, NaCl	$CH_3OCH_2CH_2OH$	$(CH_2(OH)·CH_2·NH_2)$	Zhou et al. [30]	Sol-gel
$Zn(CH_3OO)_2$ $2H_2O$	Isopropyl alcohol	$(CH_2(OH)·CH_2·NH_2)$	Chebil et al. [23]	Sol-gel
$Zn(CH_3OO)_2$ $2H_2O$, $Cu(CH_3COO)_2$	…….	Diethanolamine (DEA)	Agarwal et al. [14]	Sol-gel
$Zn(CH_3OO)_2$ $2H_2O$	2-methoxyethanol	$(CH_2(OH)·CH_2·NH_2)$	Haarindraprasad et al. [24]	Sol-gel
$Zn(CH_3OO)_2$ $2H_2O$	Dimethyl formamide	Diethanolamine (DEA)	Bhunia et al. [25]	Sol-gel
$Zn(CH_3OO)_2$ $2H_2O$, $C_2H_7NO_2$	Distilled water/glacial acetic acid	–	Para et al. [26]	Sol-gel
$Zn(CH_3OO)_2$ $2H_2O$, $Ga(NO_3)_3$ xH_2O	2-methoxyethanol	$(CH_2(OH)·CH_2·NH_2)$	Wang et al [27]	Sol-gel
[$Zn(CH_3OO)_2$ $2H_2O$	2-methoxyethanol	$(CH_2(OH)·CH_2·NH_2)$	Alfaro et al. [28]	Sol-gel
$Zn(CH_3OO)_2$ $2H_2O$, LiOH, graphene	C2H5OH/EtOH	–	Li et al. [29]	Sol-gel
$Zn(CH_3OO)_2$ $2H_2O$, $Mg(CH_3COO)_2$ $4H_2O$, $Al(NO_3)_3$ $9H_2O$)	Isopropyl alcohol	Diethanolamine (DEA)	Das et al. [13]	Sol-gel

Precursor	Solvent	Stabilizing agent	Reference	Technique
$Zn(CH_3OO)_2\ 2H_2O$	1-butanol	$(CH_2(OH)\cdot CH_2\cdot NH_2)$	Demes et al. [31]	Sol-gel
$Zn(CH_3OO)_2\ 2H_2O$, $SnCl_2.2H_2O$	Ethanol and chelating with glycerin	Acetic acid	Kose et al. [32]	Sol-gel
$Zn(CH_3OO)_2\ 2H_2O$, $Li(CH_3\text{-}COO)_2.2H_2O$,$Co(CH_3COO)_2.2H_2O$	(C_2H_5OH)	$(C_2H_6O_2)$	Bashir et al. [15]	Sol-gel
$Zn(CH_3OO)_2\ 2H_2O$	Ethanol (C_2H_5OH)	$(CH_2(OH)\cdot CH_2\cdot NH_2)$	Ayana et al. [33]	Sol-gel
$Zn(CH_3OO)_2\ 2H_2O$, $Cu(CO_2CH_3)_2\ H_2O$	Ethanol (C_2H_5OH)	$(CH_2(OH)\cdot CH_2\cdot NH_2)$	Wang et al. [16]	Sol-gel
$Zn(CH_3OO)_2\ 2H_2O$, NaOH	2-Propanol	–	Zimmermann et al. [34]	Sol-gel
$Zn(CH_3OO)_2\ 2H_2O$	Acetone	TEA	Efafi et al. [35]	Sol-gel
Zinc nitrate hexa hydrate/Na-CMC	Deionized water		Muthukrishnan et al [36].	Sol-gel
$Zn(NO_3)_2.6H_2O/Bi(NO_3)_3.5H_2O$, NaOH	Deionized water	PEG-6000	Liu et al. [37]	Sol-gel
$Ti(OCH(CH_3)_2)_4$, $Zn(CH_3COO)_2\ 2H_2O$	Isopropyl alcohol	–	Boro et al. [38]	Sol-gel
$Zn(CH_3OO)_2\ 2H_2O$, NH_4VO_3	CH_3OH/MeOH	–	Slama et al. [17]	Sol-gel
$ZnCl_2$, $FeCl_3$, NH_4Ac, $Zn(CH_3OO)_2\ 2H_2O$	$C_2H_6O_2$		Rabbani et al. [39]	Sol-gel
$(Zn(CH_3COO)_2.2H_2O)/TiO_2$	Isopropyl alcohol	$(CH_2(OH)\cdot CH_2\cdot NH_2)$	Marimuthu et al. [40]	Sol-gel
$Zn(CH_3OO)_2\ 2H_2O$, $Co(NO_3)_2.6H_2O]$	Double distilled water	$[C_6H_8O_7\ H_2O]$	Birajdar et al. [18]	Sol-gel
$Z_n(NO_3)_2$, citric acid and tetraethoxysilane	Ethanol (C_2H_5OH)	–	Sivakami et al. [41]	Sol-gel
Isopropyl orthotitanate (TTIP), zinc nitrate tetra hydrate	Ethanol (C_2H_5OH)	Diethanolamine (DEA)	Moradi et al. [42]	Sol-gel
$Zn(CH_3OO)_2\ 2H_2O$	2-Methoxyethanol	$(CH_2(OH)\cdot CH_2\cdot NH_2)$	Ocaya et al. [43]	Sol-gel
$Zn(CH_3OO)_2\ 2H_2O$, $CoCl_2$		Polyvinyl alcohol	Verma et al. [19]	Sol-gel
$[Zn(NO_3)_2\ 6H_2O]/Ga(NO_3)_3$, gelatin	Distilled water	–	Khorsand Zak et al. [20]	Sol-gel
$Zn(CH_3OO)_2\ 2H_2O$	Distilled water/ethanol	$(CH_2(OH)\cdot CH_2\cdot NH_2)$	Kiani et al. [44]	Sol-gel

Table 1. Precursors and solvents used in the synthesis of ZnO by the sol-gel process.

Precursor	Solvent	Stabilizing agent	Reference	Technique
$Zn(CH_3OO)_2$ $2H_2O$, sulfo propyl methacrylatepotassium	Ethylene glycol	–	Liua et al. [45]	Colloidal
$Zn(CH_3OO)_2$ $2H_2O$	Distilled water	Poly(vinyl alcohol) (PVA)	Nagvenkar et al. [46]	Colloidal
$Zn(CH_3OO)_2$ $2H_2O$, $LiOH{\cdot}H_2O$	Ethanol (C_2H_5OH)	–	Yuan et al. [47]	Colloidal
$Zn(CH_3OO)_2$ $2H_2O$, tetraalkylammonium hydroxide	DMSO	NEt4OH	Panasiuk et al. [48]	Colloidal
$Zn(CH_3OO)_2$ $2H_2O$	Ethanol	Triethylamine, diethylamine	Gupta et al. [49]	Colloidal
$(Zn(NO_3)_2$ $6H_2O)$, NaOH	Distilled water	1-Thioglycerol (TG) and 2 mercaptoethanol (ME)	Hodlur et al. [50]	Colloidal
$Zn(CH_3OO)_2$ $2H_2O$	Deionized water	Hexamethyl netetramine	Guo et al. [56]	Colloidal
$Zn(CH_3OO)_2$ $2H_2O$, KOH	Methanol	–	Rahman [51]	Colloidal
$Zn(CH_3OO)_2$ $2H_2O$, KOH	Methanol	PVP	Gutul et al. [52]	Colloidal
$Zn(CH_3OO)_2$ $2H_2O$, KOH	Ethanol	3-aminopropyltriethoxysilane	Moghaddam et al. [53]	Colloidal
$Zn(CH_3OO)_2$ $2H_2O$, NaOH	Ethyl alcohol	–	Liu et al. [54]	Colloidal
$Zn(CH_3OO)_2$ $2H_2O$	Diethylene glycol.	–	Xie et al. [60]	Colloidal
$Zn(CH_3OO)_2$ $2H_2O$	Ethanol	LiOH	Verma et al. [61]	Colloidal
$Zn(CH_3OO)_2$ $2H_2O$, NaOH	2-propanol	–	Moghaddam et al. [64]	Microwave
GO, $Zn(NO_3)_2$, NaOH	Deionized water	–	Tian et al. [65]	Microwave
$Zn(CH_3OO)_2$ $2H_2O$, NaOH	Distilled water	Guanidinium carbonate, acetyl acetone,	Hamedani et al. [66]	Microwave
Zinc hydroxide	Distilled water	Cetyltrimethylammonium bromide	Rai et al. [67]	Microwave
$Zn(CH_3OO)_2$ $2H_2O$, NaOH, NH_4OH	Dieonized water	–	Yanga et al. [69]	Microwave
$(Zn(NO_3)_2.6H_2O)$, hydrazine hydrate	Distilled water	–	Krishnakumar et al. [70]	Microwave
$ZnSO_4{\cdot}7H_2O$, GO, NaOH	Distilled water	–	Lua et al. [71]	Microwave
$Zn(CH_3OO)_2$ $2H_2O$	Deionized water	–	Zhu et al. [72]	Microwave

Precursor	Solvent	Stabilizing agent	Reference	Technique
$ZnSO_4$, NaOH	Deionized water	–	Liu et al. [73]	Microwave
$Zn(NO_3)_2$	Deionized water	–	Rassaeia et al. [74]	Microwave
Zinc oxide, ammonium hydroxide	Deionized water	–	Lu et al. [75]	Microwave
$ZnSO_4$, NaOH	Deionized water	–	Limaye et al. [76]	Microwave
Zinc acetylacetonate monohydrate	Water	Ethoxyethanol, ethoxyethanol, and n-butoxyethanol	Schneider et al. [77]	Microwave

Table 2. Precursors and solvents used in the synthesis of ZnO by colloidal/microwave synthesis.

3.2. Recent studies and applications

Various morphologies of ZnO can be obtained from the sol-gel process including nanorods [3], inhomogeneous films [4, 5], inhomogeneous nanoparticles [6] and nanocomposites [7].

The structural effects of cation doping on ZnO nanoparticles was investigated in several studies. When doped with lithium, it was found that the concentration of Li^+ ion substitution for Zn^{2+} directly affected the XRD intensity of the (002) plane, but did not affect the grain size or crystallinity of the nanoparticles [8]. When ZnO was doped with Ca^{2+} ions, the average particle size was increased to 40–90 nm which could be attributed to the larger ionic radius of Ca^{2+} that substituted for Zn^{2+} ion sites [9]. Likewise, the average grain size reduced when a small radius ion is substituted for Zn^{2+} (0.74 Å) in the hexagonal wurtzite structure such as Co^{2+} (0.58 Å) [10]. Doping with Al^{3+} ions also showed the same tendency in reducing particle size, however, impurity phases such as Al_2O_3 and $ZnAl_2O_4$ were also observed [11]. Additionally, co-doping of ZnO with Ag^+ and Al^{3+} ions showed the formation of crystal defects due to the difference in ionic radius between Ag^+, Al^{3+} and Zn^{2+}. Crystallinity improved proportionally with increased Ag^+ doping concentration, however, lattice defects and dislocations increased with Al^{3+} substitution [12]. Further dopant studies also demonstrated that limited dopant precursor solubility provoked a random distribution of dopant throughout the product [13]. Most research about doping ZnO has resulted in improved optical and electrical properties due to improved morphology or intrinsic material properties [14–20].

Synthesis of ZnO of different morphologies without doping is also important to consider since product morphology alone can affect device properties. Without any dopant ZnO can be obtained under normal laboratory conditions with well-aligned nanorods, agglomerated nanoparticles and inhomogeneous thin films composed of nanoparticles, quantum dots, nano-wires, spheres or nano-cubes [21–44].

Colloidal synthesis technique can be utilized to obtain nanocomposites of ZnO and other materials. Nano-sheets of poly (styrene-methyl methacrylate-sulfopropyl methacrylate potassium)/ZnO nanocomposites were obtained by Liua et al. [45]. Dissolving ZnO in other materials can result in a great combination and co-application of materials such as ZnO/PVA (Polyvinyl alcohol) [46]. The same process was done to produce ZnO/TiO_2 multilayer thin films [47]. This technique allows obtaining well size-controlled nanoparticles such as those reported with use of dimethyl sulfoxide, but the author reports that the solvent and post-annealing treatment are also important factors in the crystallization process and average particle size [48].

Several authors have reported that the product morphology can be altered between flakes, hexagons, particles and flower-like morphologies by adding different surfactant material [49]. Agglomeration of ZnO nanoparticles was reduced by adding capping agents to different thiol molecules during synthesis [50]. It was demonstrated that the colloidal stability of nanoparticles can be maintained after dispersion in monoethanolamine (MEA). Also, hybrid structures can be obtained through this method like ZnO-Au reported recently [51]. Dispersion of nanomaterials could also be maintained through an additive such as poly (N-vinylpyrrolidone) which has been shown to maintain colloidal stability for more than a couple of months [52]. In the same way agglomeration of ZnO quantum dots can be prevented through a capping agent such as 3-aminopropyltriethoxysilane in order to maintain their quantum properties [53]. Stabilization of the colloidal particles ensures that particle size and shape does not change with time allowing for more repetitive results for each batch of material. Stable colloidal solutions have also been used to grow novel nanostructures on several kinds of unique substrates such as wood that can allow for new ecological applications in future [54–63].

Colloidal and sol-gel processing are both chemical techniques that can be used to easily obtain different nanomaterials; similarly, microwave-assisted synthesis can obtain similar products but has been explored very little. In microwave-assisted synthesis, most reactions take place in a short amount of time and have resulted in the synthesis of good ZnO nanostructures. The technique has obtained spherical nanoparticles that are stable in solution for up to 50 days, and can be deposited several times on a substrate without any change in its morphology. Similarly, it is possible to obtain composites such as ZnO-nanoparticles on reduced graphene oxide. Also, the morphology is highly dependent on the complexing agent where the reaction takes place or if a dopant is added, such as that reported for obtaining ZnO nanoflowers, nanorods and nanoparticles. Additionally, a research group has confirmed the formation of flower-like to rod-like nanostructures by changing the system temperature. Other works have also reported about dumbbell-shaped nanoparticles, nano-flowers, graphene-ZnO nanocomposites, straw-bundle, chrysanthemum and nanorod-based microspheres obtained under certain temperature conditions. [2, 64–78].

4. Conclusions and future directions

The techniques listed in the above paragraphs remain as the most important chemical solution-based routes to synthesize ZnO. Within the same processing method, a variety of material morphologies and properties can be obtained by subtle changes in temperature, additives, dopants or other parameters. There has been a wide range of organic and inorganic particles that have been synthesized and applied in different fields through these techniques. Investigating the effects of processing conditions on ZnO nanoparticles is still a hot topic in current research for their applications in optoelectronic and solar cell devices.

Author details

Jose Alberto Alvarado Garcia[1*], Zachary Garbe Neale[1], Antonio Arce-Plaza[2], Avelino Cortes Santiago[3] and Hector Juarez Santiesteban[4]

*Address all correspondence to: jalvarado@cinvestav.mx

1 Department of Materials Science and Engineering University of Washington, Seattle, WA, USA

2 School of Engineering and Architecture, Zacatenco Campus, National Polytechnic Institute (ESIAZ-IPN), Mexico City, Mexico

3 Faculty of chemical sciences, Benemeritous Autonomous University of Puebla, Puebla, Mexico

4 Semiconductor Devices Research Center, Benemeritous Autonomous University of Puebla, Puebla, Mexico

References

[1] Anastas PT, Warner JC. Green Chemistry: Theory and Practice. New York: Oxford University Press; 1998. p. 152

[2] Bilecka I, Elser P, Niederberger M. Kinetic and thermodynamic aspects in the microwave-assisted synthesis of ZnO nanoparticles in benzyl alcohol. ACS Nano. 2009;**3**(2):467-477. DOI: 10.1021/nn800842b

[3] Pourshaban E, Abdizadeh H, Reza Golobostanfard M. A close correlation between nucleation sites, growth and final properties of ZnO nanorod arrays: Sol-gel assisted chemical bath deposition process. Ceramics International. 2016;**42**:14721-14729. DOI: http://dx.doi.org/10.1016/j.ceramint.2016.06.098

[4] Joshi K, Rawat M, Gautam SK, Singh RG, Ramola RC. Band gap widening and narrowing in Cu-doped ZnO thin films. Journal of Alloys and Compounds. 2016;**680**:252-258. DOI: http://dx.doi.org/10.1016/j.jallcom.2016.04.093

[5] Kasar CK, Sonawane US, Bange JP, Patil DS. Blue luminescence from Ba0.05Zn0.95O nanostructure. Journal of Materials Science. 2016;**27**:8126-8130. DOI: 10.1007/s10854-016-4814-9

[6] Kumar P, Singh BK, Pal BN, Pandey PC. Correlation between structural, optical and magnetic properties of Mn-doped ZnO. Applied Physics A. 2016;**122**(8):122-740. DOI: 10.1007/s00339-016-0265-7

[7] Bahari A, Roeinfard M, Ramzannezhad A. Characteristics of Fe3O4/ZnO nanocomposite as a possible gate dielectric of nanoscale transistors in the field of cyborg. Journal of Materials Science: Materials in Electronics. 2016;**27**:9363-9369. DOI: 10.1007/s10854-016-4978-3

[8] F. Boudjouan, A. Chelouche, T. Touam, D. Djouadi, R. Mahiou, G. Chadeyron, A. Fischer, A. Boudrioua. Doping effect investigation of Li-doped nanostructured ZnO thin films prepared by sol-gel process. Journal of Materials Science: Materials in Electronics. 2016;**27**:8040-8046. DOI: 10.1007/s10854-016-4800-2

[9] Slama R, Ghoul JEl, Omri K, Houas A, Mir L El, Launay F. Effect of Ca-doping on microstructure and photocatalytic activity of ZnO nanoparticles synthesized by sol gel method. Journal of Materials Science: Materials in Electronics. 2016;**27**:7939-7946. DOI: 10.1007/s10854-016-4786-9

[10] Dhruvash, Shishodia PK. Effect of cobalt doping on ZnO thin films deposited by sol-gel method. Thin Solid Films. 2016;**612**:55-60. DOI: http://dx.doi.org/10.1016/j.tsf.2016.05.028

[11] Tabassum S, Yamasue E. Okumura H, Ishihara KN. Electrical stability of Al-doped ZnO transparent electrode prepared by sol-gel method. Applied Surface Science. 2016;**377**:355-360. DOI: http://dx.doi.org/10.1016/j.apsusc.2016.03.133

[12] Khan F, Ho Baek S, Hyun Kim J. Enhanced charge transport properties of Ag and Al co-doped ZnO. Journal of Alloys and Compounds. 2016;**682**:232-237. DOI: http://dx.doi.org/10.1016/j.jallcom.2016.04.292

[13] Das A, Guha Roy P., Dutta A, Sen S, Pramanik P, Das D, Banerjee A, Bhattacharyya A. Mg and Al co-doping of ZnO thin films: Effect on ultraviolet photoconductivity. Materials Science in Semiconductor Processing. 2016;**54**:36-41. DOI: http://dx.doi.org/10.1016/j.mssp.2016.06.018

[14] Agarwal L, Singh BK, Tripathi S, Chakrabarti P. Fabrication and characterization of Pd/Cu doped ZnO/Si and Ni/Cu doped ZnO/Si Schottky diodes. Thin Solid Films. 2016;**612**:259-266. DOI: http://dx.doi.org/10.1016/j.tsf.2016.06.027

[15] Bashir MI, Ali K, Sarfraz AK, Mirza IM. Room temperature synthesis and multiferroic response of Li co-doped (Zn, Co)O nanocrystallites. Journal of Alloys and Compounds. 2016;**684**:151-161. DOI: http://dx.doi.org/10.1016/j.jallcom.2016.04.019

[16] Wang M, Ji J, Luo S, Jiang L, Ma J, Xie X, Ping Y, Ge J. Sol-gel-derived ZnO/Cu/ZnO multilayer thin films and their optical and electrical properties. Materials Science in Semiconductor Processing. 2016;**51**:55-59. DOI:http://dx.doi.org/10.1016/j.mssp.2016.04.020

[17] Slama R, El Ghoul J, Ghiloufi I, Omri K, El Mir L, Houas A. Synthesis and physico-chemical studies of vanadium doped zinc oxide nanoparticles and its photocatalysis. Journal of Materials Science: Materials in Electronics. 2016;**27**:8146-8153. DOI: 10.1007/s10854-016-4817-6

[18] Birajdar SD, Khirade PP, Bhagwat VR, Humbe AV, Jadhav KM. Synthesis, structural, morphological, optical and magnetic properties of $Zn_{1-x}Co_xO$ ($0 \leq x \leq 0.36$) nanoparticles synthesized by sol-gel auto combustion method. Journal of Alloys and Compounds. 2016;**683**:513e526. DOI: http://dx.doi.org/10.1016/j.jallcom.2016.05.043

[19] Verma KC, Bhatia R, Kumar S, Kotnala RK. Vacancies driven magnetic ordering in ZnO nanoparticles due to low concentrated Co ions. Materials Research Express. 2016;**3**:076103. DOI: doi:10.1088/2053-1591/3/7/076103

[20] Khorsand Zak A, Suhail Abd Aziz N, Manaf Hashim A, Kordi F. XPS and UV–vis studies of Ga-doped zinc oxide nanoparticles synthesized by gelatin based sol-gel approach. Ceramics International. 2016;**42**:13605-13611. DOI: http://dx.doi.org/10.1016/j.ceramint.2016.05.155

[21] Singh M, Yusuf Mulla M, Vittoria Santacroce M, Magliulo M, Di Franco C, Manoli K, Altamura D, Giannini C, Cioffi N, Palazzo G, Scamarcio G, Torsi L. Effect of the gate metal work function on water-gated ZnO thin-film transistor performance. Journal of Physics D: Applied Physics. 2016;**49**:275101. DOI: doi:10.1088/0022-3727/49/27/275101

[22] Kim OS, Kang BH, Lee JS, Lee SW, Cha SH, Lee JW, Kim SW, Kim SH, Kang SW. Efficient quantum dots light-emitting devices using polyvinyl pyrrolidone-capped ZnO nanoparticles with enhanced charge transport. IEEE Electron Device Letters. 2016;**37**(8):1022-1024. DOI: 10.1109/LED.2016.2578304

[23] Chebil W, Boukadhaba MA, Fouzri A. Epitaxial growth of ZnO on quartz substrate by sol-gel spinccoating method. Superlattices and Microstructures. 2016;**95**:48-55. DOI: http://dx.doi.org/10.1016/j.spmi.2016.04.033

[24] Haarindraprasad R, Hashim U, Gopinath SCB, Perumal V, Liu WW, Balakrishnan SR. Fabrication of interdigitated high-performance zinc oxide nanowire modified electrodes for glucose sensing. Analytica Chimica Acta. 2016;**925**:70-81. DOI: http://dx.doi.org/10.1016/j.aca.2016.04.030

[25] Bhunia R, Das S, Dalui S, Hussain S, Paul R, Kumar Pal RBA. Flexible nano-ZnO/polyvinylidene difluoride piezoelectric composite films as energy harvester. Applied Physics A. 2016;**122**:122-637. DOI: 10.1007/s00339-016-0161-1

[26] Ahmad Para T, Ahmad Reshi H, Pillai S, Shelke V. Grain size disposed structural, optical and polarization tuning in ZnO. Applied Physics A. 2016;**122**:122-730. DOI: 10.1007/s00339-016-0256-8

[27] Wang H, Sun Y, Fang L, Wang L, Chang B, Sun X, Ye L. Growth and characterization of high transmittance GZO films prepared by sol-gel method. Thin Solid Films. 2016;**615**:19-24. DOI: http://dx.doi.org/10.1016/j.tsf.2016.06.048

[28] Alfaro Cruz MR, Hernandez-Como N, Mejia I, Ortega-Zarzosa G, Martınez-Castañon G-A, Quevedo-Lopez MA. Impact of the annealing atmosphere in the electrical and optical properties of ZnO thin films. Journal of Sol-Gel Science and Technology. 2016;**79**:184-189. DOI: 10.1007/s10971-016-4035-y

[29] Li H, Wei Y, Zhang Y, Zhang C, Wang G, Zhao Y, Yin F, Bakenov Z. In situ sol-gel synthesis of ultrafine ZnO nanocrystals anchored on grapheneas anode material for lithium-ion batteries. Ceramics International. 2016;**42**:12371-12377. DOI: http://dx.doi.org/10.1016/j.ceramint.2016.05.010

[30] Zhou M, iZang D, Zhai X, Gao Z, Zhang W, Wang C. Preparation of biomorphic porous zinc oxide by wood template method. Ceramics International. 2016;**42**:10704-10710. DOI: http://dx.doi.org/10.1016/j.ceramint.2016.03.188

[31] Demes T, Ternon C, Riassetto D, Roussel H, Rapenne L, Gélard I, Jimenez C, Stambouli V, Langlet M. New insights in the structural and morphological properties of sol-gel deposited ZnO multilayer films. Journal of Physics and Chemistry of Solids. 2016;**95**:43-55. DOI: http://dx.doi.org/10.1016/j.jpcs.2016.03.017

[32] Kose H, Dombaycıoglu S, Osman Aydın A, Akbulut H. Production and characterization of free-standing ZnO/SnO2/MWCNT ternary nanocomposite Li-ion battery anode. International Journal of Hydrogen Energy. 2016;**41**:9924-9932. DOI: http://dx.doi.org/10.1016/j.ijhydene.2016.03.202

[33] Gemechu Ayana D, Ceccato R, Collini C, Lorenzelli L, Prusakova V, Dirè S. Sol-gel derived oriented multilayer ZnO thin films with memristive response. Thin Solid Films. 20166;**615**:427-436. DOI: http://dx.doi.org/10.1016/j.tsf.2016.07.025

[34] Zimmermann LM, Baldissera P, Bechtold IH. Stability of ZnO quantum dots tuned by controlled addition of ethylene glycol during their growth. Materials Research Express. 2016;**3**:075018. DOI: 10.1088/2053-1591/3/7/075018

[35] Efafi B, Majles Ara MH, Mousavi SS. Strong blue emission from ZnO nanocrystals synthesized in acetone-based solvent. Journal of Luminescence. 2016;**178**:384-387. DOI: http://dx.doi.org/10.1016/j.jlumin.2016.06.026

[36] Muthukrishnan K, Vanaraja M, Boomadevi S, Kumar Karn R, Singh V, Singh PK, Pandiyan K. Studies on acetone sensing characteristics of ZnO thin film prepared by sol-gel dip. Journal of Alloys and Compounds. 2016;**673**:138-143. DOI: http://dx.doi.org/10.1016/j.jallcom.2016.02.222

[37] Ting Liu T, Hua Wang M, Ping Zhang H. Synthesis and characterization of ZnO/Bi_2O_3 core/shell nanoparticles by the sol-gel method. Journal of Electronic Materials. 2016;**45**(8): 4412-4417. DOI: 10.1007/s11664-016-4568-4

[38] Boro B, Rajbongshi BM, Samdarshi SK. Synthesis and fabrication of TiO_2–ZnO nanocomposite based solid state dye sensitized solar cell. Journal of Materials Science: Materials in Electronics. 2016;**27**:9929-9940. DOI: 10.1007/s10854-016-5062-8

[39] Rabbani M, Heidari-Golafzani M, Rahimi R. Synthesis of TCPP/ZnFe2O4@ZnO nanohollow sphere composite for degradation of methylene blue and 4-nitrophenol under visible light. Materials Chemistry and Physics. 2016;**179**:35-41. DOI: http://dx.doi.org/10.1016/j.matchemphys.2016.05.005

[40] Marimuthu T, Anandhan N, Thangamuthu R, Mummoorthi M, Ravi G. Synthesis of ZnO nanowire arrays on ZnO-TiO_2 mixed oxide seed layer for dye sensitized solar cell applications. Journal of Alloys and Compounds. 2016;**677**:211-218. DOI: http://dx.doi.org/10.1016/j.jallcom.2016.03.219

[41] Sivakami R, Thiyagarajan P. The effect of citric acid on morphology and photoluminescenceproperties of white light emitting ZnO-SiO_2 nanocomposites. Photonics and Nanostructures—Fundamentals and Applications. 2016;**20**:31-40. DOI: http://dx.doi.org/10.1016/j.photonics.2016.03.003

[42] Moradi S, Aberoomand-Azar P, Raeis-Farshid S, Abedini-Khorrami S, Hadi Givianrad M. The effect of different molar ratios of ZnO on characterization and photocatalytic activity of TiO_2/ZnO nanocomposite. Journal of Saudi Chemical Society. 2012;**20**:373-378. DOI: http://dx.doi.org/10.1016/j.jscs.2012.08.002

[43] Ocaya RO, Al-Ghamdi A, El-Tantawy F, Farooq WA, Yakuphanoglu F. Thermal sensor based zinc oxide diode for low temperature applications. Journal of Alloys and Compounds. 2016;**674**:277-288. DOI: http://dx.doi.org/10.1016/j.jallcom.2016.02.267

[44] Kiani A, Dastafkan K. Zinc oxide nanocubes as a destructive nanoadsorbent for the neutralization chemistry of 2-chloroethyl phenyl sulfide: A sulfur mustard simulant. Journal of Colloid and Interface Science. 2016;**478**:271-279. DOI: http://dx.doi.org/10.1016/j.jcis.2016.06.025

[45] Liua J, Hub ZY, Penga Y, Huanga HW, Lia Y, Wua M, Keb XX, Van Tendeloob G, Su BL. 2D ZnO mesoporous single-crystal nanosheets with exposed {0001} polar facets for the depollution of cationic dye molecules by highly selective adsorption and

photocatalytic decomposition. Applied Catalysis B: Environmental. 2016;**181**:138-145. DOI: http://dx.doi.org/10.1016/j.apcatb.2015.07.054

[46] Nagvenkar AP, Deokar A, Perelshteina I, Gedanken A. A one-step sonochemical synthesis of stable ZnO–PVA nanocolloid as a potential biocidal agent. Journal of Materials Chemistry B. 2016;**4**:2124. DOI: 10.1039/c6tb00033a

[47] Yuan S, Mu J, Mao R, Li Y, Zhang Q, Wang H. All-Nanoparticle Self-assembly ZnO/TiO_2 Heterojunction Thin Films with Remarkably Enhanced Photoelectrochemical Activity. ACS Applied Materials & Interfaces. 2014;**6**:5719–5725. DOI: dx.doi.org/10.1021/am500314n

[48] Panasiuk YV, Raevskaya OE, Stroyuk OL, Kuchmiy SY, Dzhagan VM, Hietschold M, Zahn DRT. Colloidal ZnO nanocrystals in dimethylsulfoxide: A new synthesis, optical, photo- and electroluminescent properties. Nanotechnology. 2014;**25**:075601. DOI: 10.1088/0957-4484/25/7/075601

[49] Gupta A, Srivastava A, Bahadur L, Amalnerkar DP, Chauhan R. Comparison of physical and electrochemical properties of ZnO prepared via different surfactant-assisted precipitation routes. Applied Nanoscience. 2015;**5**:787-794. DOI: 10.1007/s13204-014-0379-1

[50] Hodlur RM, Rabinal MK, Mohamed Ikram I. Influence of dipole moment of capping molecules on the optoelectronic properties of ZnO nanoparticles. Journal of Luminescence. 2014;**149**:317-324. DOI: http://dx.doi.org/10.1016/j.jlumin.2014.01.055

[51] Rahman DS, Kumar Ghosh S. Manipulating electron transfer in hybrid ZnO–Au nanostructures: Size of gold matters. Journal of Physical Chemistry. 2016;**120**:14906–14917. DOI: 10.1021/acs.jpcc.6b03551

[52] Gutul T, Rusu E, Condur N, Ursaki V, Goncearenco E, Vlazan P. Preparation of poly(N-vinylpyrrolidone)-stabilized ZnO colloid nanoparticles. Beilstein Journal of Nanotechnology. 2014;**5**:402-406. DOI: doi:10.3762/bjnano.5.47

[53] Moghaddam E, Youzbashi AA, Kazemzadeh A, Eshraghi MJ. Preparation of surface-modified ZnO quantum dots through anultrasound assisted sol-gel process. Applied Surface Science. 2015;**346**:111-114. DOI: http://dx.doi.org/10.1016/j.apsusc.2015.03.207

[54] Liu Y, Fu Y, Yu H, Liu Y. Process of in situ forming well-aligned Zinc Oxide nanorod arrays on wood substrate using a two-step bottom-up method. Journal of Colloid and Interface Science. 2013;**407**:116-121. DOI: http://dx.doi.org/10.1016/j.jcis.2013.06.043

[55] Jhaveri JH, Murthy ZVP. A comprehensive review on anti-fouling nanocomposite membranes for pressure driven membrane separation processes. Desalination. 2016;**379**:137-154. DOI: http://dx.doi.org/10.1016/j.desal.2015.11.009

[56] Guo D, Sato K, Hibino S, Takeuchi T, Bessho H, Kato K. Low-temperature preparation of transparent conductive Al-doped ZnO thin films by a novel solgel method. Journal of Materials Science. 2014;**4**:4722-4734. DOI: 10.1007/s10853-014-8172-9

[57] Koushkia E, Farzaneha A, Majles Arab MH. Modeling absorption spectrum and saturation intensity of ZnO nano-colloid. Optik. 2014;**125**:220-223. DOI: http://dx.doi.org/10.1016/j.ijleo.2013.06.007

[58] Senthil Kumar K, Chandramohan S, Natarajan P. Photophysics and photochemistry of phenosafranine adsorbed on the surface of ZnO loaded nanoporous materials. Dyes and Pigments. 2014;**109**:206-213. DOI: http://dx.doi.org/10.1016/j.dyepig.2014.05.008

[59] Gwaka GH, Leeb WJ, Paekb SM, Oha JM,. Physico-chemical changes of ZnO nanoparticles with different size and surface chemistry under physiological pH conditions. Colloids and Surfaces B: Biointerfaces. 2015;**127**:137-142. DOI: http://dx.doi.org/10.1016/j.colsurfb.2015.01.021

[60] Xie J, Wang H, Duan M. QCM chemical sensor based on ZnO colloid spheres for the alcohols. Sensors and Actuators B. 2014;**203**:239-244. DOI: http://dx.doi.org/10.1016/j.snb.2014.06.119

[61] Verma SD, Sharma SN, Kharkwal A, Bhagavannarayana G, Kumara M, Nath Singha S, Kumar Singha P, Sazad Mehdibb S, Husain M. Role of nanocrystalline ZnO coating on the stability of porous silicon formed on textured (100) Si. Applied Surface Science. 2013;**285p**:564-571. DOI: http://dx.doi.org/10.1016/j.apsusc.2013.08.094

[62] Lin W, Haderlein M, Walter J, Peukert W, Segets D. Spectra library: An assumption-free in situ method to access the kinetics of catechols binding to colloidal ZnO quantum dots. Angewandte Chemie International Edition. 2016;**55**:932-935. DOI: 10.1002/anie.201508252

[63] Zhang G, Morikawa H, Chen Y. Synthesis of ZnO nanoparticles in aqueous solution and their antibacterial activities. Japanese Journal of Applied Physics. 2014;**53**:06JG07. DOI: http://dx.doi.org/10.7567/JJAP.53.06JG07

[64] Moghaddam FM, Saeidian H. Controlled microwave-assisted synthesis of ZnO nanopowder and its catalytic activity for O-acylation of alcohol and phenol. Materials Science and Engineering B. 2007;**139**:265-269. DOI: 10.1016/j.mseb.2007.03.002

[65] Tian L, Pan L, Liu X, Lu T, Zhu G, Sun Z. Enhanced photocatalytic degradation of methylene blue by ZnO-reduced graphene oxide composite synthesized via microwave-assisted reaction. Journal of Alloys and Compounds. 2011;**509**:10086-10091. DOI: 10.1016/j.jallcom.2011.08.045

[66] Hamedania NF, Mahjouba AR, Ali Khodadadib A, Mortazavi Y. Microwave assisted fast synthesis of various ZnO morphologies for selective detection of CO, CH4 and ethanol. Sensors and Actuators B. 2011;**156**:737-742. DOI: 10.1016/j.snb.2011.02.028

[67] Rai P, Song HM, Kim Yun-Su, Song MK, Oh PR, Yoon JM, Yu YT. Microwave assisted hydrothermal synthesis of single crystalline ZnO nanorods for gas sensor application. Materials Letters. 2012;**68**:90-93. DOI: 10.1016/j.matlet.2011.10.029

[68] Kim YJ, Varma RS. Microwave-assisted preparation of cyclic ureas from diamines in the presence of ZnO. Tetrahedron Letters. 2004;**45**:7205-7208. DOI: 10.1016/j.tetlet.2004.08.042

[69] Yanga LY, Donga SY, Suna JH, Fenga JL, Wua QH, Sun S-P. Microwave-assisted preparation, characterization and photocatalytic properties of a dumbbell-shaped ZnO photocatalyst. Journal of Hazardous Materials. 2010;**179**:438-443. DOI: 10.1016/j.jhazmat.2010.03.023

[70] Krishnakumar T, Jayaprakash R, Pinna N, Singh VN, Mehta BR, Phani AR. Microwave-assisted synthesis and characterization of flower shaped zinc oxide nanostructures. Materials Letters. 2009;**63**:242-245. DOI: 10.1016/j.matlet.2008.10.008

[71] Lua T, Pana L, Li H, Zhua G, Lva T, Liua X, Suna Z, Chenb T, Chuab DHC. Microwave-assisted synthesis of graphene–ZnO nanocomposite for electrochemical supercapacitors. Journal of Alloys and Compounds. 2011;**509**:5488-5492. DOI: 10.1016/j.jallcom.2011.02.136

[72] Zhu P, Zhang J, Wu Z, Zhang Z. Microwave-assisted synthesis of various ZnO hierarchical nanostructures: Effects of heating parameters of microwave oven. Crystal Growth & Design. 2008;**8**(9): 3148-3153. DOI: 10.1021/cg0704504

[73] Xinjuan Liu, Likun Pan, Tian Lv, Ting Lu, Guang Zhu, Zhuo Sun and Changqing Sun. Microwave[M5]-assisted synthesis of ZnO–graphene composite for photocatalytic reduction of Cr(VI). Catalysis Science & Technology 2011;**1**:1189-1193. DOI: 10.1039/c1cy00109d

[74] Rassaeia L, Jabera R, Flowera SE, Edlera KJ, Comptonb RG, Jamesa TD, Marken F. Microwave-electrochemical formation of colloidal zinc oxide at fluorine doped tin oxide electrodes. Electrochimica Acta. 2010;**55**:7909-7915. DOI: 10.1016/j.electacta.2010.01.068

[75] Lu CH, wang WJH, Godbole SV. Microwave-hydrothermal synthesis and photoluminescence characteristics of zinc oxide powders. Journal of Materials Research. 2005;**20**(2):464. DOI: 10.1557/JMR.2005.0067

[76] Limaye MV, Singh S, Das R, Poddar P, Kulkarni SK. Room temperature ferromagnetism in undoped and Fe doped ZnO nanorods: Microwave-assisted synthesis. Journal of Solid State Chemistry. 2011;**184**:391-400. DOI: 10.1016/j.jssc.2010.11.008

[77] Schneider JJ, Hoffmann RC, Engstler J, Klyszcz A, Erdem E, Jakes P, Eichel RA, Bauermann LP, Bill J. Synthesis, characterization, defect chemistry, and FET properties of microwave-derived nanoscaled zinc oxide. Chemistry of Materials. 2010;**22**:2203-2212. DOI: 10.1021/cm902300q

[78] Guo Y, Wang H, He C, Qiu L, Cao X. Uniform carbon-coated ZnO nanorods: Microwave-assisted preparation, cytotoxicity, and photocatalytic activity. Langmuir. 2009;**25**(8):4678-4684. DOI: 10.1021/la803530h

Chapter 2

Ferrite Nanostructures Consolidated by Spark Plasma Sintering (SPS)

Romain Breitwieser, Ulises Acevedo,
Souad Ammar and Raul Valenzuela

Additional information is available at the end of the chapter

http://dx.doi.org/10.5772/68017

Abstract

Ferrites are a well-known class of ferrimagnetic materials. In the form of nanoparticles (NPs), they exhibit novel and fascinating properties, leading to an extremely wide variety of applications in electronics, biomedical and environmental fields. These applications result from nanoscale effects on physical properties, particularly magnetic properties. For applications in electronic devices, however, a high-density, consolidated body, with very fine grains is needed, in order to retain the nanoscale properties. To our knowledge, spark plasma sintering (SPS) is the only method permitting a full densification with final grain size in the nanometer range. In this review, we examine the SPS method as applied to ferrites and, in particular, the effects of SPS parameters on the final nanostructures obtained. Due to their technological impact, we also discuss the SPS fabrication of hybrid multiferroic nanostructures composed of a ferrite and a ferroelectric phase.

Keywords: spark plasma sintering, ferrites, nanostructures

1. Introduction

Ferrite materials have been widely used because of their excellent magnetic and electric properties. With three basic crystal structures, spinel, garnet and hexagonal unit cells, they can form a virtually endlessly number of solid solutions and thus be tailored for most applications [1]. Ferrite-based devices [2] are used in all communication systems (magnetic storage, microwave absorbers, high frequency inductors, radar devices, phase array antennas), different electric motors (a modern car has more than 30 DC motors), as permanent magnets. Hexagonal ferrites still constitute about 75% of the permanent magnet world market [3].

Developed since the 1930s, ferrites are sometimes considered as a “mature” technology, with the implication that advances in this field would only be incremental. This, in fact, is very far from reality [4]. A simple look at specialized literature shows that advances in ferrite processing and devices during the last 10 years have been dramatic and significant.

The ferrite size reduction from bulk to nanoscale has opened a whole new field of potential applications, as well as many issues on basic research. Magnetic properties of nanoparticles can be quite different [5] from those of their microsized or bulk counterparts. As the dimensions are reduced below a critical size, magnetic properties change from those of multidomain to those of single domain structure. Magnetic moments at the particle surface become noncollinear due to broken exchange bonds at the external layer of the particles. An additional complexity arises from the interactions (dipolar, exchange) between particles, a subject which so far, has been poorly understood. The macroscopic magnetic properties of nanoparticle-based materials can be therefore extremely complex. But as it occurs very often, a source of complexity can become also a source of novel applications and knowledge once properly understood.

Nanoferrite technology has been ongoing for the last 20 years, with recent advances in magnetic fine particle production by many innovative routes. However, for most applications in electronic devices, a nanoparticle powder is unsuitable; a high density and fully consolidated material is needed. Conventional sintering–usually operated at temperatures in excess of 1100°C for a few hours–leads to an excessive grain growth, making such a method unacceptable to consolidate nanoscale materials.

A novel sintering technique has been recently developed in Japan, known as spark plasma sintering (SPS) [6], consisting of generating internally heat by means of DC current pulses directly through a graphite die containing the powder compact. This technique allows very fast heating and cooling rates (up to 1000°C/min) and has shown its potential to consolidate nanoparticle powders into high-density samples with controlled grain growth. Contrary to conventional sintering, ferrites have been consolidated to high density at temperatures as low as 500°C with sintering times as short as 5 min, while keeping grain size in the 100 nm range [7].

Recently, SPS has also shown [8, 9] the capability to consolidate nanohetero-structures in order to fabricate multifunctional materials of strong scientific and technological interest. Typically, ferrites combined with ferroelectric perovskites may lead to extrinsic multiferroics in which the magnetization of the former can be driven by the application of an electric field and vice-versa. The efficiency of this magnetoelectric exchange depends critically on the morphology of interfaces [10, 11], which in turn is controlled by the SPS process parameters.

In this chapter, we present a review of nanostructures that have been obtained by means of SPS. In Section 2, a self-contained review of the basic aspects of ferrites is given. Section 3 describes the main features of the spark plasma sintering method. The methods to synthesize ferrite NPs that can be used as precursors for the subsequent SPS consolidation are briefly described in Section 4. A review of results obtained on spinel ferrites using SPS is given in Section 5. The consolidation of ferrites with garnet and hexagonal unit cells is discussed in Sections 6 and 7, respectively. Section 8 describes the application of SPS to the fabrication of ferrite-based multiferroic composites, and a general conclusion is given in Section 9.

2. Overview of crystal structure and magnetic properties of ferrites

2.1. Spinels

Spinel ferrites possess the crystal structure of the natural spinel $MgAl_2O_4$, where Mg and Al have been substituted by transition cations. This structure is particularly stable, and there is a large variety of oxides which adopt it, fulfilling the conditions of overall cation-to-anion ratio of 3/4, a total cation valence of 8. An important ferrite is magnetite, $Fe^{2+}Fe^{3+}{}_2O_4(Fe_3O_4)$, the oldest known magnetic solid which is still being researched due to the fascinating properties associated with the coexistence of ferrous and ferric cations. Another important ferrite material is maghemite or γ-Fe_2O_3, which is a defective spinel $Fe^{3+}{}_{8/3}\square_{1/3}O_4$, where □ represents vacancies on octahedral cation sites. The cation distribution, that is, the occupancy of sites by cations depends on many parameters [1] such as ion size, valence, electronic symmetry, and even cooling rate when they are prepared by sintering, which is the most frequently used process. When divalent cations occupy tetrahedral (or "A") sites and trivalent ones are on octahedral (or "B") sites, it is known as a normal spinel. A different cation distribution, where half of trivalent cations occupy A sites and B sites, is shared by divalent cations and the remaining trivalent cations are known as the inverse spinel. An intermediate cation distribution has been obtained for some ferrites, where cations are statistically distributed on the two sites.

A remarkable characteristic of spinel structure is that it is able to form an extremely wide variety of solid solutions by the partial or total substitution of divalent, or trivalent cations, leading to an extremely wide diversity of magnetic properties. Magnetic transition temperature, for instance, varies from 9 K for $ZnFe_2O_4$ to –958 K for $Li_{0.5}Fe_{2.5}O_4$ [1].

2.2. Garnets

Magnetic garnets possess the crystal structure of mineral $Mn_3Al_2Si_3O_{12}$, with rare-earth (RE) and Fe^{3+} cations instead, leading to the general formula $RE_3Fe_5O_{12}$, RE is in the series from La^{3+} to Lu^{3+}. The garnet unit cell includes 160 atoms; rare-earth cations occupy dodecahedral sites, and iron cations are distributed among the tetrahedral (three) and octahedral (two) sites. An additional case is yttrium iron garnet, known as "YIG", $Y_3Fe_5O_{12}$, which can be prepared as a very homogeneous phase and exhibits remarkable magnetic properties, especially at microwave frequencies.

As in the case of spinels, cations can be partially or totally substituted to obtain a large number of compositions. Also, as in spinels, magnetic interactions are of the superexchange type. Rare-earth and octahedral magnetic sublattices are oriented parallel and opposite to tetrahedral sublattice. The spontaneous magnetic moments in rare-earth cations are localized in orbital 4*f*, which is an internal one; additionally, dodecahedral sites are very large. These two factors lead to a diminished superexchange interaction, which decreases rapidly with increasing temperature. The main results are that all rare-earth magnetic garnets exhibit virtually the same Curie temperature (560–573 K), as well as a "compensation temperature" where magnetization vanishes due to the mutual compensation of sublattice magnetic moments, and which depends on the rare-earth cation [1].

2.3. Hexaferrites

Hexaferrites are a "family" of hexagonal (and some rhombohedral) closely related structures. The main formulae can be written as follows: $BaFe_{12}O_{19}$ (*M*), $Ba_2Me_2Fe_{12}O_{22}$ (*Y*), $BaMe_2Fe_{16}O_{27}$ (*W*), $Ba_2Me_2Fe_{28}O_{46}$ (*X*), $Ba_4Me_2Fe_{36}O_{60}$ (*U*) and $Ba_6Me_{24}Fe_{48}O_{82}$ (*Z*). Me represents a divalent cation such as Zn, Cu, Mg, Co and Mn. Ba can also be substituted totally or partially by Ca, Sr, Pb. Hexagonal ferrites lead to an extremely wide variety of compositions and hence magnetic properties. Their magnetic structures are quite complex; for example, in the *M* structure, the magnetization results from interaction between five sublattices coupled by superexchange interactions. Hexaferrites are used mostly as permanent magnets due to their hard magnetic properties.

2.4. Magnetic properties of ferrite nanoparticles

The nanoscale adds a complexity level to the magnetic properties of materials [12]. Most of the critical lengths for magnetic properties and structures fall in the nanometer range. Two of the most important changes are the transition from multidomain to single domain and then the change from the long-range ordered ferrimagnetic to the superparamagnetic (SPM) behavior. The former occurs when a nanoparticle (NP) becomes very small, and the energy to produce a domain wall is larger than the magnetostatic energy produced by a single domain configuration. A single domain structure is usually characterized by a maximum on the coercive field. If the size of the particles is still decreased, then the coercive field and the remanent magnetization can vanish as SPM state takes place. Superparamagnetism [5] appears when the particle is very small, and its total anisotropy energy E_K ($E_K = K{\cdot}V$, with K = anisotropy constant and V = particle volume) is smaller than the thermal energy $E_T = k_B{\cdot}T$ (with k_B = Boltzmann constant and T = temperature). Thermal energy thus overcomes anisotropy, and the magnetization oscillates randomly. The main difference with the classic paramagnetic behavior is that in the case of SPM, it is the whole magnetization vector which exhibits random reversals, instead of the individual spins. Magnetization, that is, spins coupling is maintained up to the Curie point, T_C.

Since anisotropy energy usually *decreases* with temperature, and thermal energy *increases* with T, there exists a "blocking" temperature, T_B, for which $E_K = E_T$; below this point NPs behave as ordered ferromagnetic (anisotropy dominates), and above T_B, NPs become superparamagnetic, with random magnetization.

Superparamagnetism is catastrophic for materials intended for magnetic recording, since any information (in the form of a particular magnetic domain structure) is lost at room temperature. For medical applications, however, SPM is just the magnetic behavior needed. In cancer tumor removal techniques, magnetic NPs (conveniently functionalized) are injected into the patient to be selectively localized onto the tumor. Single domain ferromagnetic NPs, due to their magnetostatic fields, can become aggregated and form large clusters. Once the tumor is eliminated, magnetic NPs can be easily recovered with a magnet because under the effect of a magnetic field, SPM NPs acquire a significant magnetization and are attracted.

The consolidation of NPs by SPS leads to the formation of a polycrystalline structure with grains separated by grain boundaries and the tendency to grain growth. Both of them strongly decrease the surface effects observed in NPs. Magnetic properties can therefore exhibit dramatic changes as compared with the properties of original NPs powder; typically, saturation magnetization increases to approach the bulk value, and coercive field shows a definite value (and thus a remanent magnetization). An interesting case is cobalt ferrite, which can be synthesized as fine NPs in the 5–10 nm range [13]. For such small NPs, cobalt ferrite shows a SPM behavior at room temperature, as in **Figure 1(a)**. Zero-field-cooled, field-cooled (ZFC-FC) experiments usually lead to a blocking temperature in the 100–150 K range (**Figure 1(b)**). Below this T_B, a normal ferromagnetic behavior is observed; at higher temperatures, the anisotropy is overwhelmed by thermal agitation and a SPM phase is observed. However, when these NPs are consolidated by SPS, a typical hysteresis loops with well-defined coercivity and remanence are observed at all temperatures (below the Curie point), as shown in **Figure 1(c)** [14].

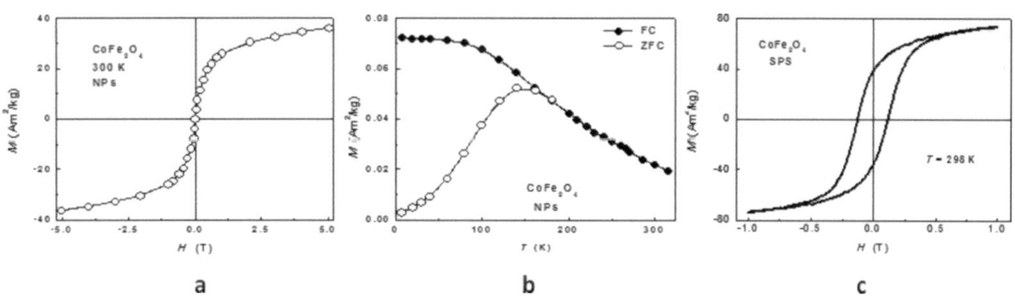

Figure 1. (a) SP behavior of cobalt ferrite NPs in the 10 nm diameter range; (b) magnetization behavior in ZFC-FC measurements, showing a blocking temperature ca. 150 K; and (c) magnetic hysteresis of consolidated sample at 500°C for 5 min (adapted from Refs. [13, 14]).

3. The spark plasma sintering process

In the SPS process, the sample (typically a powder) is placed in graphite die and then compressed by two pistons, **Figure 2**. The heating is by Joule effect, by electric current pulses. High currents (up to 1500 A) and strong pressures (up to 120 MPa with graphite dies and up to 600 MPa with tungsten carbide dies) can be applied, which allow significantly large heating rates (up to 1000°C/min). It is a versatile technique with many processing parameters: sintering temperature T_s, heating and cooling rates (*Hr*, *Cr*, respectively), maximum load (P_s), load rate *Pr*, load removal rate (*Pc*). By measuring the distance between pistons (Z) and its more sensitive time derivative (d*Z*/d*t*) during the process, an estimation of the density changes is recorded, which has a direct relationship with sintering. A typical temperature and shrinkage profile are shown in **Figure 2**. The main objective of the process is to transform a particle powder into a high density solid. In some cases, a chemical reaction can occur during the process, in addition to densification; in this review, only the densification process is discussed.

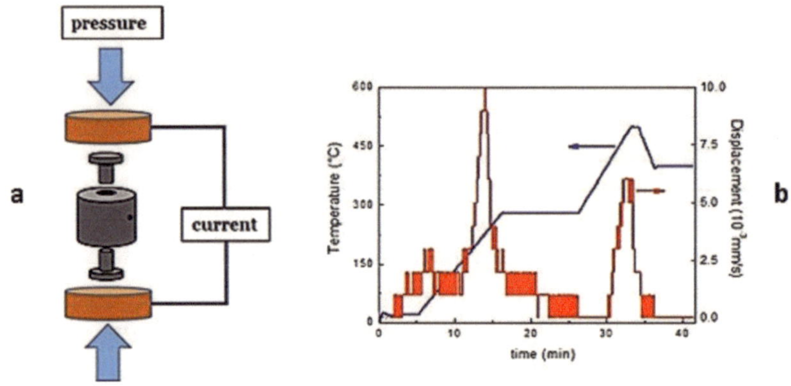

Figure 2. (a) SPS electrodes and die set up; (b) a typical temperature profile (left axis), and shrinkage trace (right axis), in the form of time derivative d*Z*/d*t* for a spinel ferrite (adapted from Ref. [7]).

The densification process, that is, the removal of porosity (corresponding to empty spaces between the original particles) requires diffusion; this is produced, of course, by increasing the temperature. It appears, however, that electric current pulses have an effect also on atomic diffusion, since high densities can be reached at very low temperatures and extremely short times as compared with classic sintering processes. This additional effect of current has been attributed to an increase in point defect concentration [15], as well as a reduction in the activation energy for mobility of defects [16]. In the case of insulators, it appears that it is the electric field that leads to such enhanced diffusion since these materials can also be effectively consolidated by SPS.

The heating rate can have a strong effect on the final microstructure of the material. During the process, there is a competition between densification and grain growth (or "coarsening"). The former, which is basically the removal of porosity, depends on volume diffusion, while grain growth is controlled essentially by surface diffusion. Volume diffusion involves mass transport and hence occurs at higher temperatures than surface diffusion. A high heating rate therefore can "bypass" the range of grain growth alone and start readily the densification with limited coarsening. The possibility to separate these two mechanisms makes SPS the ideal, and so far, the only suitable process to fabricate full density nanostructured materials.

The pressure has a significant influence on final density. At relatively low temperatures, an increase in pressure leads to higher densities. At high temperatures, the increase in density is accompanied by an increase also in grain size. A measure of the shrinkage rate [17] can be obtained from the time derivative of distance between pistons, d*Z*/d*t*, which is directly obtained in most SPS available systems.

4. Brief review of synthesis methods for ferrite nanopowders

To date, a variety of synthetic methods have been applied to produce magnetic nanoparticles, including those of ferrites. These methods can be divided into two main approaches, a top down and a bottom up ones. Bottom-up routes are attractive in terms of their low cost and sustainability; there is, however, a generic challenge in directly obtaining particles with a calibrated size and a uniform shape. They mainly consist on a direct precipitation of the desired phase starting from selected precursors dissolved for an adjusted atomic ratio in a given solvent and heated at a high enough temperature. The involved reactions depend on the operating conditions: hydrolysis [9, 13], thermal decomposition [18, 19] and condensation.

In some cases, the synthesis is indirect and an intermediate solid phase, like an alcoholate, a hydroxide or a hydroxycarboxylate, is first obtained and then calcined at moderated heating conditions (less than 800°C for a couple of hours in air) in order to form the target phase (see for instance [20, 21]).

The heating conditions of the reaction solution in all these wet chemical methods can be conventional or microwave assisted [22, 23]. They can be achieved under standard pressure or under autogenous pressure (solvothermal route [24–27]). The synthesis can be achieved under high intensity ultrasounds (sonochemical technique [28]) or under energetic radiation (radiolysis method [29, 30]). Surfactants or polymers can be introduced in the reaction medium to improve the control of the size and the shape of the particles and/or to avoid their aggregation.

The top-down method is based on an energetic milling process of a powder [31]. The powder can be the commercial bulk ferrite phase or a mixture of the raw oxide or hydroxides of the metallic cations constituting its composition. The milling process can be carried out using different apparatus, namely attritor, planetary mill or a linear ball milling (**Figure 3**) but in all the cases, the introduced powders are cold welded and fractured during mechanical alloying, favoring in one case its nanostructuration and its chemical reactivity [32, 33].

For all nanocrystalline materials prepared by high-energy ball milling synthesis route, less-effective size control, surface and interface contamination and amorphization are major concerns. In particular, mechanically attributed contamination by the milling tools (Fe, WC, etc.) is often the problem. Recently, surfactant-assisted ball milling has been exploited for the synthesis of various nanomaterials, nanograins, and nanocomposites from solid bulk materials.

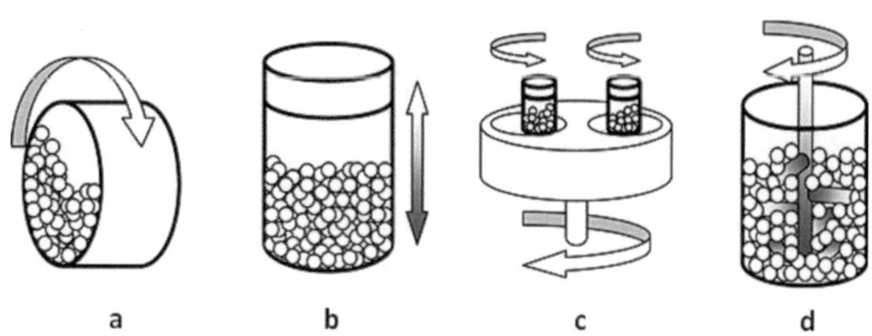

Figure 3. Schematic representation of (a) the indirect direct milling equipment, including (b) linear and (c) planetary mill equipment and that of the direct one (d), called attritor. Republished with permission of the Royal Society of Chemistry, from Ref. [34]; permission conveyed through Copyright Clearance Center, Inc.

Promising results have been obtained in terms of nanostructure assembly, micro-/nanoenvironment, surface functionalities, and dispersion [35].

Focusing on spinel ferrites, all these approaches were successfully used to produce nanoparticles. For instance, cobalt ferrite particles having different sizes were prepared from 2 to 50 nm by sol-gel method [20] and from 2 to 10 nm by polyol process [7, 9, 13, 14, 36], without using any additional capping agents/surfactants. In these cases, the particle size was varied by controlling the temperature of the calcination of the precipitated intermediate solid and that of the heating reaction solution, respectively. As this temperature is elevated, the size of the resulting particles is large, and their crystalline quality is high (see for instance [37–40]). Moreover, such nanoparticles can exhibit a deviation from the thermodynamic spinel stable structure. This deviation, mainly consisting on a change in the cation occupancy of spinel sites, increases with temperature.

By opposition, garnets and hexaferrites nanopowders can be scarcely produced directly. Whatever the employed synthesis route, wet chemistry or ball-milling, the total processing route often involves a subsequent annealing step in order to obtain highly crystallized and pure perovskite or hexagonal phases. For instance, nanostructured M-type $SrFe_{12}O_{19}$ hexaferrite powder was produced by combining forced hydrolysis in polyol followed by subsequent annealing at 836°C [41]. It was also successfully prepared by combining ball milling followed by annealing at 700°C [42]. This annealing step makes the resulting nanopowders typically polycrystalline of several tens of nanometers in crystal size with a magnetization relatively close to that measured on their bulk counterparts. The situation is almost the same for nanostructured garnets. Annealing is required after ball-milling [43] or after sol-gel precipitation [44] for YIG preparation. In some cases, the precursors from these reactions can be used not only for SPS sintering but also for "reactive" SPS treatment, which include the chemical reaction.

5. Consolidation of spinel ferrites by SPS

In a typical SPS powder consolidation experiment, the starting powder is put in the electrically conductive toroidal die made of graphite, and uniaxially pressed by the upper and lower graphite punches. Most of the electric current flows through the mould, and specimen is heated up by Joule effect. The electric current can also flow through the powder materials itself, and self-heating occurs at the inside of the sample. Low temperature and short time sintering are therefore possible. In the case of spinel ferrites, this technique was first used to reduce sintering time on micrometer-sized powders [45, 46]. Then, it was used to reduce the sintering temperature using nanometer-sized powders [7, 9, 47–49].

Spinel nanoferrites were readily synthesized by soft chemistry methods (see Section 3) in a wide variety of compositions and can be directly consolidated by SPS. Ni-Zn ferrites have been consolidated at temperatures as low as 350°C for 10 min, to densities about 85% [50]. Density and grain size increased by increasing the sintering temperature to 400 and 500°C, as appears in **Figure 4**.

http://dx.doi.org/10.5772/68017

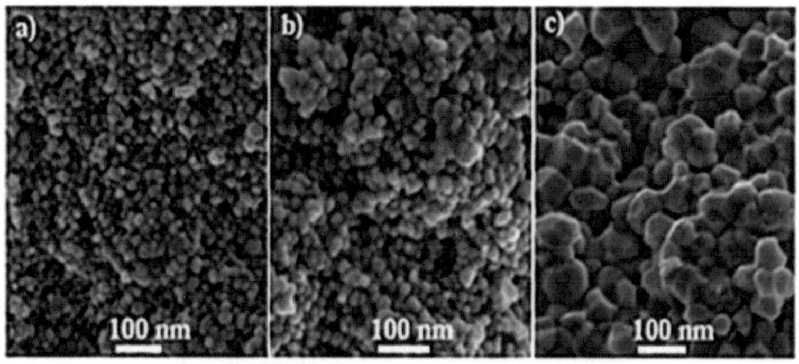

Figure 4. Nanostructures produced by consolidation of Ni-Zn NPs by SPS by 10 min at (a) 350°C, (b) 400°C, and (c) 500°C [51].

Magnetite can be consolidated by SPS into a high density, high crystallinity solid at 750°C for 15 min, with a final grain size in the 150 nm range [49], as shown in **Figure 5(a)**. The reductive atmosphere of SPS prevented the oxidation of initial magnetite NPs to maghemite, γ-Fe_2O_3, first, and then to hematite, α-Fe_2O_3. The quality of the crystals can be assessed by means of the resolution of the Verwey transition [52], which occurs about 120 K in two steps: first the crystallographic transition [53] from the monoclinic to the cubic system and then the spin reorientation from uniaxial to cubic symmetry [54], as shown in **Figure 5**.

The aggregation state of initial NPs can have an important effect on the final grain size of the consolidated sample (**Figure 6**). By varying the polyol during the NPs synthesis (by hydrolysis in a polyol method), different aggregation clustering can be achieved [9], from monodispersed NPs (reaction with diethylenglycol), ~10 NPs clusters (by using propanediol), and ~100 NPs clusters (prepared with ethanediol).

The size of the NPs was the same (~5 nm) in all clusters. The final grain sizes after consolidation exhibited the opposite tendency, that is, the largest grain size was found for the monodispersed sample. This points to the dependence of grain growth on surface diffusion; large

a

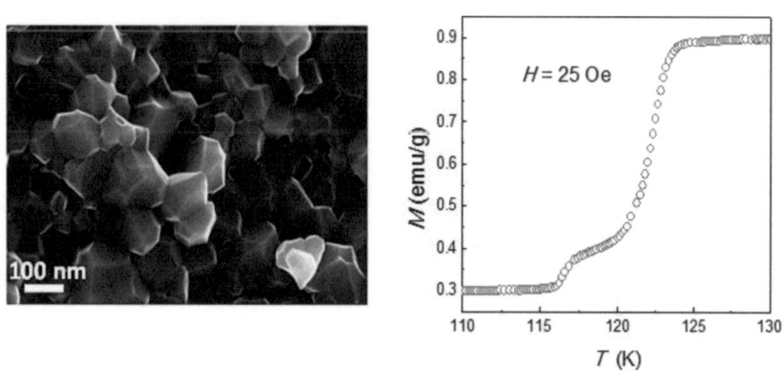

b

Figure 5. (a) SEM micrograph of magnetite consolidated by SPS at 750°C, 15 min and (b) magnetization measurements showing the Verwey transition (adapted from Ref. [49]).

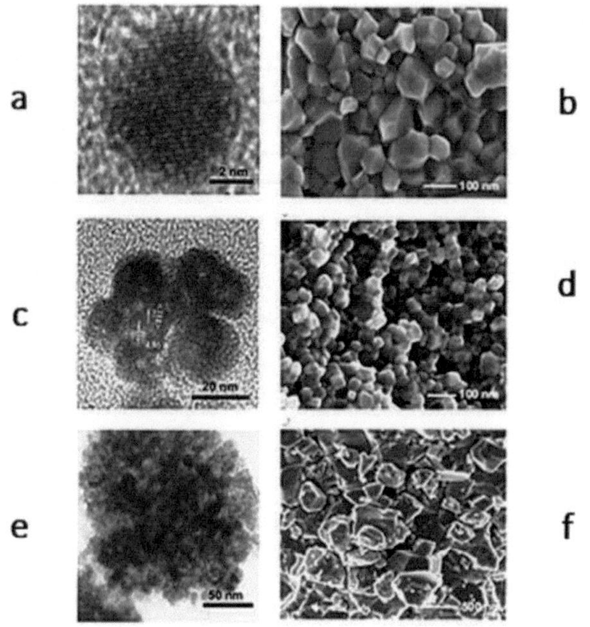

Figure 6. Polyol-synthesized NPs in different media: (a) diethylenglycol, (c) propanediol and (e) ethanediol; they led to SPS consolidated nanostructures (b), (d) and (f), respectively. Reprinted from Ref. [9] Copyright (2014), with permission from Elsevier.

clusters grow mainly through the cluster surface, while monodispersed NPs coalesce directly from their individual surface.

The importance of heating rate appears in **Figure 7**, where the nanostructure of a cobalt ferrite is shown for SPS processing at the same sintering temperature and times (2 min at 600°C followed by 5 min at 500°C), but with heating rates of 80°C/min, and 15°C/min [9]. Grain size is about 70 nm for the former, and 290 nm for the slow heating rate. This illustrates again the differences between surface and volume diffusion, as previously discussed.

Figure 7. Co ferrite processed with different heating rates to reach the sintering temperature (600°C): left 80°C/min, and right 15°C/min. Reprinted from Ref. [9] Copyright (2014), with permission from Elsevier.

http://dx.doi.org/10.5772/68017

6. Consolidation of garnets

To the best of our knowledge, only hydrothermal, and more generally solvothermal, synthesis techniques may allow the more or less direct production of ferrites with the garnet structure. From a general point of view, dense oxide structures, namely crystalline cells containing a high number of atoms per unit volume like orthoferrite, garnet or hexaferrite cells, require a high activation energy to nucleate. This can be achieved by a postannealing treatment as explained above or by increasing the pressure during reaction solution heating.

If a consolidated phase is sought, then a SPS treatment can be applied by using the resulting precursors of a soft chemistry synthesis reaction. Yttrium iron garnet (YIG) can be prepared by a SPS treatment at 750°C for 15 min at 100 MPa from the precursors obtained from a polyol synthesis reaction [55], in big contrast with the typical parameters of the solid state reaction (1350°C for a few hours). Note that in these original shaping conditions, highly dense and ultrafine grained ceramics (**Figure 8**) were systematically obtained.

Figure 8. SEM micrograph of YIG obtained by polyol reaction + SPS treatment at 750°C, 15 min, 100 MPa. Reprinted with permission from Ref. [55], Copyright (2013) Wiley.

7. Synthesis of hexaferrites by SPS

Nanostructured hexaferrites can be synthesized by various methods such as sol-gel [56], coprecipitation [57], spray-drying and microemulsion [58], ball-milling [42], among others. However, the obtained products still need thermal posttreatments in the range of 800–1200°C to crystallize into hexagonal ferrites and get sintered. Different ways are used to minimize the grain growth before sintering. The annealed particles can be postmilled, resulting in smaller particles that in turn increase the density and decrease the sintering temperatures [59]. Another way is the containment of the particles in an amorphous glassy matrix before their annealing to form the desired phase while limiting its grain growth. One of the most impressive result is the crystallization of 57 nm M-ferrites obtained at 642°C with the addition of B_2O_3 glass vs 260 nm at 800°C with silica [60]. Besides, doping hexaferrites with Zn^{2+} and La^{3+} can also act as a grain growth inhibitor during annealing treatment [61].

Recently, the spark plasma sintering (SPS) method has brought an interesting alternative to synthesis/sinter hexaferrites and optimize their magnetic properties. Indeed, SPS is carried out at relatively low temperatures in a very short time of few minutes only, and additionally, it allows a convenient control of grain growth. Such observations apply to SrM-type hexaferrites prepared by the mechanical compounding method [62]. The resulting ceramics have a fine microstructure as illustrated in **Figure 9** [63].

SPS of $SrFe_{12}O_{19}$ at 1100°C for 5 min leads to a maximum density of 5.15 Mg/m^3, while the maximum density obtained by ordinary sintering at 1240°C for 2 h is only 4.83 Mg/m^3. In this case, the samples prepared by SPS showed half-sized grains (400 nm) and harder magnetic properties. The effect of SPS temperature and time on the density and magnetic properties of SrM ferrites was also significant. A high energy product $(BH)_{max}$ of 18.3 kJ/m^3 was found for such strontium hexaferrites [63]. Similarly, the sintering of Ba hexaferrites by SPS has also led to harder magnetic properties due to the limited grain growth: grains of ~100–150 nm by SPS vs 1.5–8 μm by conventional sintering [64].

SPS is not only useful to consolidate nanopowders, but it also promotes the chemical reaction of nanoparticles. Thus, SPS has been also used to complete the formation of nanostructured hexaferrites. SPS treatments were applied to a high-energy ball milled mixture of SrO and Fe_2O_3 precursors. The formation of 70 nm M-type strontium hexaferrites has been achieved at only 700°C, under 800 MPa [65]. XRD measurements revealed that such hexagonal ferrites crystallize from nucleation sites of $Fe_2Sr_2O_5$ that is an intermediate metastable phase formed at 600°C. In comparison with the simple annealing of the same ball milled precursors at 800°C, denser SPS treated samples show higher saturation magnetization (67 emu/g vs 58 emu/g) and lower coercivity (3.7 kOe vs 6.4 kOe) (**Figure 10(a)**) that are interesting features

Figure 9. Ball-milled $SrFe_{12}O_{19}$ NPs annealed at different temperatures: (a) 700 and (c) 900°C and their related ceramics (b and d, respectively) resulting from SPS consolidation. Reprinted with permission from Ref. [63] Copyright (2016) American Chemical Society.

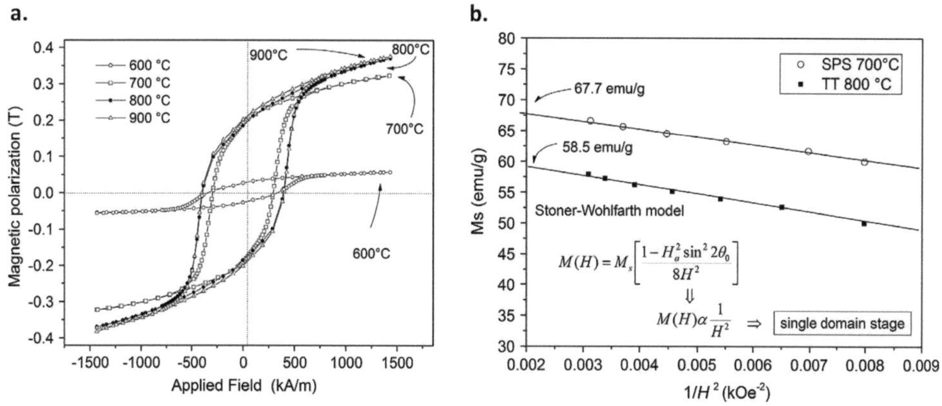

Figure 10. Sr-$Fe_{12}O_{19}$ prepared by SPS-assisted mecanosynthesis: (a) Hysteresis loops *M* vs *H* showing the effect of the SPS temperature on the magnetic properties and (b) comparison of the saturated magnetization *Ms* vs $1/H^2$ obtained after either SPS heating or conventional annealing. Republished with permission from Ref. [65] Copyright 2014, with permission through Copyright Clearance Center, Inc., Elsevier.

for recording media and electromagnetic fields wave absorption [66]. The linear M vs $1/H^2$ relationship–well fitted with the Stoner-Wohlfarth model of noninteracting single domains with uniaxial anisotropy–indicates that the decrease in coercivity is most likely due to the grain growth reduction below the single domain limit (**Figure 10(b)**). This result therefore contrasts with SPS-induced harder magnetic properties discussed earlier [62, 64], since reducing the grain growth in a range above or below the single domain limit has opposite effects on the coercivity.

The SPS-assisted synthesis of barium hexaferrites has also been demonstrated. $BaFe_{12}O_{19}$ was synthesized and sintered by SPS operated at 800°C for 10 min [67]. Interestingly, textured samples in nanobelt and nanorod microstructures were obtained showing an aspect ratio of 200 nm/several μm. Compared to textureless barium hexaferrites, the SPS-induced microstructures are characterized by higher Ms (65.5 emu/g) and lower coercivity H_c (1.4 kOe) reflecting an easier magnetic domain wall movement and/or a coherent magnetization rotation often observed for uniform microsized multidomains. It also reveals that the ultra-fast decomposition of $BaCO_3$ and $Fe(OH)_3$ precursors by SPS results in volume shrinkage leading to an accommodated nanostructure of barium hexaferrite. Furthermore, SPS can also produce hexaferrite bulk magnets of highly aligned single-magnetic domains without applying any magnetic field. This has been reported for the SPS compaction of $SrFe_{12}O_{19}$ nanoplatelets synthesized by supercritical hydrothermal flow method [68]. This effect is attributed to the highly aligned crystallites of the as-synthesized nanoplatelets together with the oriented crystal growth during SPS along the c-axis. Thus, SPS compacted samples at 950°C under 80 MPa for 2 min are composed of 80–100 nm nanocrystallites characterized by single-crystal-like magnetic properties, that is, high coercivity H_c (301 kA m^{-1}), M_s (69 A m^2 Kg^{-1}), M_r (59 A m^2 Kg^{-1}) and energy product *BH* (26 kJ m^{-3}).

As a final note, the SPS heating and pressing not only permits to decrease the crystallization/sintering temperatures of hexaferrites and to tailor their magnetic properties, but it also prevents the formation of secondary phases such as α-Fe_2O_3 [68]. On the other hand, the reductive atmosphere in SPS can lead to hexaferrite decomposition and the subsequent magnetite formation (reduction of Fe^{3+} to Fe^{2+}) [69]. Nevertheless, a recent study has shown that such a Fe_3O_4 secondary phase can be suppressed by using protecting discs of aluminum oxide between the graphite mold and the samples [66].

8. Preparation of hybrid multiferroic composites by SPS

Multiferroic materials are commonly understood as those materials exhibiting at least two ferroic orders (i.e., ferroelectric and ferro- or ferrimagnetic orders) coupled in the same matrix, leading to a magneto-electric effect (ME), that is, a magnetic field induces a change in the electric response, or vice-versa. A strong interest has been aroused by such materials since many high-tech applications can be based on it. While ME effects may exist in a single phase material, the combination of two phases (i.e., a ferromagnet and a ferroelectric) in the same matrix (a "hybrid" multiferroic) has many advantages [70]. The ME effect can be tuned by varying the phase composition and interphase connectivity, which depend directly from the nanostructure.

The ME effects in hybrid multiferroics can be enhanced by increasing the contact area between the involved phases [10, 11]. In the case of homogeneously mixed grained composites, an effective way to do this is retaining the grain growth during the sintering process in order to keep the final grain size at nanoscale. It will increase the grain boundary surface and therefore the contact interface between different phases. The SPS technique has been demonstrated to be a very fast and efficient route to consolidate high-density bulk oxide ceramics, retaining grain sizes of the sintered materials at the scale of ~100 nm [7, 9, 14]. A very small grain size means a larger interphase surface and, therefore, a stronger ME coupling.

8.1. Spinel ferrite-based composites

The ferrite (ferromagnetic)/perovskite (ferroelectric) system has been mainly investigated. Nickel ferrite $NiFe_2O_4$/lead zirconate titanate PZT (or NFO/PZT for short) was investigated by Jiang from laboratory synthesized and commercial submicrometer sized precursors, respectively [71]. They obtained layered and homogeneously mixed multiferroic composites with submicrometer sized grains by using sintering temperatures between 900 and 1050°C, and pressures between 50 and 100 MPa (**Figure 11**). A certain degree of interphase diffusion between both phases was reported for this system. Although the last system was found to be suitable for practical applications because of their good ME properties and high density, the employment of lead in ceramics and other materials is nowadays discarded by most of the scientific community because of their impact on the environment.

To solve the lead problem, barium or strontium titanate is used, together with cobalt ferrite. In fact, the system $CFO_x/BTO_{(1-x)}$ has been widely researched. Hybrid multiferroics in this system have been prepared from commercial nanopowder reagents [72], commercial BTO and autocombustion synthesized CFO [73], at temperatures in the 1050°C range. A reduction in temperature (860°C) [74] was not enough to eliminate the interphase diffusion. The use

http://dx.doi.org/10.5772/68017

Figure 11. SEM image of homogeneous $NiFe_2O_4$ (NFO) ferrite/PZT composite revealing the grain size and morphology; the inset shows the interface between ferrite (NFO) and PZT (adapted from Ref. [71].

of starting NPs synthesized by coprecipitation and ball milling allowed [75] to eliminate the interdiffusion and improve the ME coupling coefficient.

The use of both initial NPs synthesized by soft chemistry, low temperatures in the 650°C, higher pressures (100 Mpa) and short sintering times (5 min), producing grain sizes in the 150 nm (see **Figure 12**), led to high ME coefficients and the absence of interdiffusion [76]. The use of Impedance Spectroscopy (i.e., the analysis of the electrical response of the material in a wide frequency range) exhibited three main components which can be associated [77] with the three corresponding interfaces: BTO-BTO, CFO-CFO, and CFO-BTO. Experiments as a function of frequency, temperature and applied magnetic field allowed establishing the polarization processes in the three interfaces.

In another work, Co-substituted Ni-Zn ferrite was synthesized using a nonconventional solid state reaction method and sintered by SPS at 850°C and 100 MPa [79]. This ferrite was used to make layered composites by pasting a layer of commercial PZT to the sintered pellet with silver epoxy. ME performances obtained with an optimized composition of NiCoZn ferrite/PZT bilayer were compared to those obtained with the same structure combining Terfenol-D and PZT. It was shown that the low piezomagnetic coupling performance of the ferrite (in comparison to the Terfenol-D) is balanced by a low compliance. Furthermore, they used Co-substituted Ni-Zn ferrite associated with PZT in a tri-layer structure to design a ME current sensor. This sensor can be readily used in the field of power electronic applications.

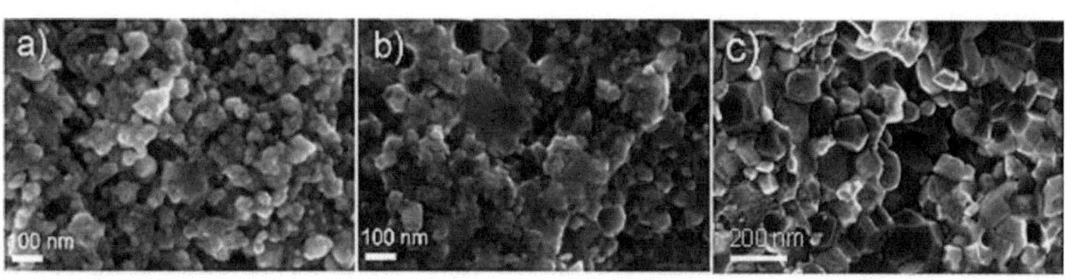

Figure 12. SEM micrographs of $CFO_x/BTO_{(1-x)}$ composites with (a) $x = 0.25$, (b) $x = 0.75$ and (c) $x = 0.50$, sintered at 650°C and 100 MPa for 5 min [78].

8.2. Other hybrid multiferroic composites

Additional efforts in this area have been reported recently in other type of ferrite-based hybrid multiferroics sintered by SPS. Bismuth ferrite (BFO) and BTO composites, sintered by SPS at temperatures around 800°C, and uniaxial pressures of 50 MPa have been prepared [80] from precursors with an average particle size of 1 μm leading to final microstructures with grain sizes all above 2 μm (**Figure 13**). Strontium hexaferrite and BTO-based hybrid multiferroics by SPS, starting from nanosized precursors purchased commercially, were investigated [81]. They were fabricated by SPS at a temperature of 900°C, a uniaxial pressure of 75 MPa and a sintering time of 3 min. These conditions (low T and short time) allowed final microstructures with grain sizes down to 200 nm and no interphase diffusion.

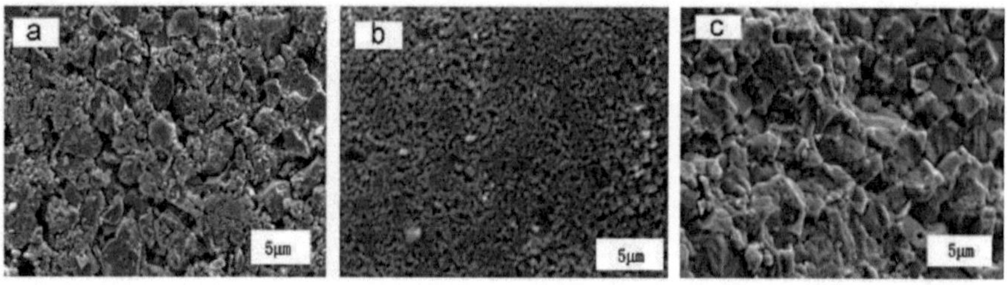

Figure 13. SEM images of 75% BFO-25%BTO at (a) 30 min mechanochemical synthesis, (b) 48 h of mechanochemical synthesis, and (c) after SPS sintering at 1000°C. Reprinted from Ref. [80] Copyright (2012), with permission from Elsevier [80].

8.3. General trends

The general trends in fabrication and characterization of hybrid multiferroics by SPS can be summarized as follows. It is possible to fabricate new high-density hybrid multiferroics by SPS, with enhanced dielectric, magnetic, and ME properties. Important variables such as the interphase diffusion, final microstructure of the sintered composites, as well as the nature and microstructural features of the starting precursors, all play a crucial role on the design of these materials. A fine grained high-density microstructure, with homogeneous average grain sizes down to 100 nm and well-defined grain boundaries, can be produced from soft-chemistry synthesized NPs. In order to retain the already formed phases of NPs and avoid interphase diffusion, short time and low temperature sintering is necessary. Sintering temperatures near to or higher than 1000°C are not recommended because a possible promotion of the interphase diffusion could result in a decrease of the dielectric constant.

9. Conclusions

At present, SPS appears to be the only practical method to produce high-density solids with grain size in the nanometer range. Additionally, SPS requires relatively lower temperatures and shorter times for densification compared to conventional sintering. In a very general way, final density increases with temperature, and grain size increases with time, so there is a compromise. The solution can be to use a relatively high sintering temperature, with a high heating rate. Finally, a problem with oxides can be the strong reductive medium of SPS, which can lead to reduction to metallic impurities in an oxide. Again, this can be minimized by means of short sintering times.

Acknowledgements

RV acknowledges a sabbatical fellowship from DGAPA-UNAM Mexico.

Author details

Romain Breitwieser[1], Ulises Acevedo[1,2], Souad Ammar[1] and Raul Valenzuela[1,2*]

*Address all correspondence to: monjaras@unam.mx

1 ITODYS, Paris Diderot University, Sorbonne Paris Cité, Paris, France

2 Materials Research Institute, National Autonomous University of Mexico, Mexico City, Mexico

References

[1] Valenzuela R. Magnetic Ceramics. UK: Cambridge University Press; 2005. 312 p.

[2] Snelling EC. Soft Ferrites: Properties and Applications. USA: Ltd, Butterworth-Heinemann; 1988. 376 p.

[3] Vukovich DP. Overview of the World Magnet Supply: Llc, Alliance; 2013.

[4] Harris VG, Geiler A, Chen Y, Yoon SD, Wu M, Yang A, et al. Recent advances in processing and applications of microwave ferrites. Journal of Magnetism and Magnetic Materials. 2009;**321**(14):2035-47.

[5] Guimaraes AP. Principles of Nanomagnetism. Berlin: Springer, Verlag; 2009. 221 p.

[6] Guillon O, Gonzalez-Julian J, Dargatz B, Kessel T, Schierning G, Rathel J, et al. Field-assisted sintering technology/spark plasma sintering: Mechanisms, materials, and technology developments. Advanced Engineering Materials. 2014;**16**(7):830-849.

[7] Valenzuela R, Beji Z, Herbst F, Ammar S. Ferromagnetic resonance behavior of spark plasma sintered Ni-Zn ferrite nanoparticles produced by a chemical route. Journal of Applied Physics. 2011;**109**(7):07A329.

[8] Nygren Z, Shen M. Microstructural prototyping of ceramics by kinetic engineering: Applications of spark plasma sintering. Chemical Record. 2005;**5**(3):173-184.

[9] Gaudisson T, Artus M, Acevedo U, Herbst F, Nowak S, Valenzuela R, et al. On the microstructural and magnetic properties of fine-grained $CoFe_2O_4$ ceramics produced by combining polyol process and spark plasma sintering. Journal of Magnetism and Magnetic Materials. 2014;**370**:87-95.

[10] Ma J, Hu J, Li Z, Nan CW. Recent progress in multiferroic magnetoelectric composites: From bulk to thin films. Advanced Materials. 2011;**23**(9):1062-1087.

[11] Schileo G. Recent developments in ceramic multiferroic composites based on core/shell and other heterostructures obtained by sol–gel routes. Progress in Solid State Chemistry. 2013;**41**(4):87-98.

[12] Gubin SP, Koksarov YA, Khomutov GB, Yurkov GY. Magnetic nanoparticles: Preparation, structure and properties. Russian Chemical Reviews. 2005;**74**(6):489-520

[13] Artus M, Tahar LB, Herbst F, Smiri L, Villain, Yaacoub FN, et al. Size-dependent magnetic properties of $CoFe_2O_4$ nanoparticles prepared in polyol. Journal of Physics: Condensed Matter. 2011;**23**(50):506001.

[14] Acevedo U, Gaudisson T, Ortega-Zempoalteca R, Nowak S, Ammar S, Valenzuela R. Magnetic properties of ferrite-titanate nanostructured composites synthesized by the polyol method and consolidated by spark plasma sintering. Journal of Applied Physics. 2013;**113**(17):17B519.

[15] Munir Z, Anselmi-Tamburini U, Ohyanagi M. The effect of electric field and pressure on the synthesis and consolidation of materials: A review of the spark plasma sintering method. Journal of Materials Science. 2006;**41**(3):763-777.

[16] Garay J, Glade S, Anselmi-Tamburini U, Asoka-Kumar P, Munir Z. Electric current enhanced defect mobility in Ni_3Ti intermetallics. Applied Physics Letters. 2004;**85**(4):573-575.

[17] Shen Z, Johnsson M, Zhao Z, Nygren M. Spark plasma sintering of alumina. Journal of the American Ceramic Society. 2002;**85**(8):1921-1927.

[18] Yu W, Falkner J, Yavuz C, Colvin V. Synthesis of monodisperse iron oxide nanocrystals by thermal decomposition of iron carboxylate salts. Chemical Communications. 2004(20):2306-2307.

[19] Soundararajan D, Kim K. Synthesis of $CoFe_2O$ magnetic nanoparticles by thermal decomposition. Journal of Magnetics. 2014;**19**(1):5-9.

[20] Naseri M, Saion E, Ahangar H, Shaari A, Hashim M. Simple synthesis and characterization of cobalt ferrite nanoparticles by a thermal treatment method. Journal of Nanomaterials. 2010;**2010**: 907686-94 .

[21] Chinie AM, Stefan A, Georgescu S. Synthesis by a citrate sol-gel method and characterization of Eu^{3+} doped yttrium aluminium garnet nanocrystals. Romanian Reports in Physics. 2005;**57**(3):412-417.

[22] Lai Z, Xu G, Yalin Z. Microwave assisted low temperature synthesis of MnZn ferrite nanoparticles. Nanoscale Research Letters. 2007;**2**(1):40-43.

[23] Kalyani S, Sangeetha J, Philip J. Microwave assisted synthesis of ferrite nanoparticles: Effect of reaction temperature on particle size and magnetic properties. Journal of Nanoscience and Nanotechnology. 2015;**15**(8):5768-5774.

[24] Daou T, Pourroy G, Begin-Colin S, Greneche J, Ulhaq-Bouillet C, Legare P, et al. Hydrothermal synthesis of monodisperse magnetite nanoparticles. Chemistry of Materials. 2006;**18**(18):4399-4404.

[25] Koseoglu Y, Alan F, Tan M, Yilgin R, Ozturk M. Low temperature hydrothermal synthesis and characterization of Mn doped cobalt ferrite nanoparticles. Ceramics International. 2012;**38**(5):3625-3634.

[26] Wang J, Ren F, Yi R, Yan A, Qiu G, Liu X. Solvothermal synthesis and magnetic properties of size-controlled nickel ferrite nanoparticles. Journal of Alloys and Compounds. 2009;**479**(1-2):791-796.

[27] Yanez-Vilar S, Sanchez-Andujar M, Gomez-Aguirre C, Mira J, Senaris-Rodriguez M, Castro-Garcia S. A simple solvothermal synthesis of MFe_2O_4 (M = Mn, Co and Ni) nanoparticles. Journal of Solid State Chemistry. 2009;**182**(10):2685-2690.

[28] Xu H, Zeiger B, Suslick K. Sonochemical synthesis of nanomaterials. Chemical Society Reviews. 2013;**42**(7):2555-2567.

[29] Abedini A, Daud A, Hamid M, Othman N, Saion E. A review on radiation-induced nucleation and growth of colloidal metallic nanoparticles. Nanoscale Research Letters. 2013;**8:1-10**.

[30] Abedini A, Daud A, Hamid M, Othman N. Radiolytic formation of Fe_3O_4 nanoparticles: Influence of radiation dose on structure and magnetic properties. Plos One. 2014;**9**(3):e90055.

[31] Yadav TP, Yadav RM, Singh DP. Mechanical milling: A top down approach for the synthesis of nanomaterials and nanocomposites. Journal of Nanoscience and Nanotechnology. 2012;**2**(3):22-48.

[32] Gaffet E, Abdellaoui M, Malhourous-Gaffet N. Formation of nanostructured materials induced by mechanical processings (overview). Materials Transactions JIM. 1995;**36**:198-209.

[33] Tan O, Cao W, Hu Y, Zhu W. Nanostructured oxides by high-energy ball milling technique: Application as gas sensing materials. Solid State Ionics. 2004;**172**(1-4):309-316.

[34] Gorrasi G, Sorrentino A. Mechanical milling as a technology to produce structural and functional bio-nanocomposites. Green Chemistry. 2015;**17**(5):2610-2625.

[35] Ullah M, Ali M, Abd Hamid S. Surfactant-assisted ball milling: A novel route to novel materials with controlled nanostructure—A review. Reviews on Advanced Materials Science. 2014;**37**(1-2):1-14.

[36] Cruz-Franco B, Gaudisson T, Ammar S, Bolarin-Miro A, Sanchez de Jesus F, Mazaleyrat F, et al. Magnetic properties of nanostructured spinel ferrites. IEEE Transactions on Magnetics. 2014;**50**(4):2800106.

[37] Ammar S, Helfen A, Jouini N, Fievet F, Rosenman I, Villain F, et al. Magnetic properties of ultrafine cobalt ferrite particles synthesized by hydrolysis in a polyol medium. Journal of Materials Chemistry. 2001;**11**(1):186-92.

[38] Beji Z, Smiri L, Yaacoub N, Greneche J, Menguy N, Ammar S, et al. Annealing effect on the magnetic properties of polyol-made Ni-Zn ferrite nanoparticles. Chemistry of Materials. 2010;**22**(4):1350-66.

[39] Ammar S, Jouini N, Fievet F, Beji Z, Smiri L, Moline P, et al. Magnetic properties of zinc ferrite nanoparticles synthesized by hydrolysis in a polyol medium. Journal of Physics-Condensed Matter. 2006;**18**(39):9055-9069.

[40] Hu P, Yang H, Pan D, Wang H, Tian J, Zhang S, et al. Heat treatment effects on microstructure and magnetic properties of Mn-Zn ferrite powders. Journal of Magnetism and Magnetic Materials. 2010;**322**(1):173-177.

[41] Gonzalez F, Miro A, De Jesus F, Escobedo C, Ammar S. Mechanism and microstructural evolution of polyol mediated synthesis of nanostructured M-type $SrFe_{12}O_{19}$. Journal of Magnetism and Magnetic Materials. 2016;**407**:188-94.

[42] Sanchez-De Jesus F, Bolarin-Miro A, Cortes-Escobedo C, Valenzuela R, Ammar S. Mechanosynthesis, crystal structure and magnetic characterization of M-type $SrFe_{12}O_{19}$. Ceramics International. 2014;**40**(3):4033-4038.

[43] Sanchez-De Jesus F, Cortes C, Valenzuela R, Ammar S, Boarin-Miro A. Synthesis of $Y_3Fe_5O_{12}$ (YIG) assisted by high-energy ball milling. Ceramics International. 2012;**38**(6):5257-5263.

[44] Wang M, Zhu X, Wei X, Zhang L, Yao X. Preparation and annealing process of $Y_3Fe_5O_{12}$ by sol-gel method. Ferroelectrics. 2001;**264**(1-4):1907-1912.

[45] Yamamoto S, Horie S, Tanamachi N, Kurisu H, Matsuura M. Fabrication of high-permeability ferrite by spark-plasma-sintering method. Journal of Magnetism and Magnetic Materials. 2001;**235**(1-3):218-222.

[46] Sun J, Li J, Sun G, Qu W. Synthesis of dense NiZn ferrites by spark plasma sintering. Ceramics International. 2002;**28**(8):855-888.

[47] Gaudisson T, Beji Z, Herbst F, Nowak S, Ammar S, Valenzuela R. Ultrafine grained high density manganese zinc ferrite produced using polyol process assisted by spark plasma sintering. Journal of Magnetism and Magnetic Materials. 2015;**387**:90-95.

[48] Valenzuela R, Ammar S, Nowak S, Vazquez G. Low field microwave absorption in nanostructured ferrite ceramics consolidated by spark plasma sintering. Journal of Superconductivity and Novel Magnetism. 2012;**25**(7):2389-2393.

[49] Gaudisson T, Vazquez-Victorio G, Banobre-Lopez M, Nowak S, Rivas J, Ammar S, et al. The Verwey transition in nanostructured magnetite produced by a combination of chimie douce and spark plasma sintering. Journal of Applied Physics. 2014;**115**(17):17E117.

[50] Ortega-Zempoalteca R, Flores-Arias Y, Vazquez-Victorio G, Gaudisson T, Ammar S, Vargas-Osorio Z, et al. The effects of spark plasma sintering consolidation on the ferromagnetic resonance specta (FMR) of Ni-Zn ferrites. Physica Status Solidi A. 2014;**211**:1062-1066.

[51] Ammar S et al, Unpublished results.

[52] Verwey EJW. Electronic conduction of magnetite (Fe_3O_4) and its transition point at low temperatures. Nature. 1939;**144**:327-328.

[53] Senn MS, Wright JP, Attfield JP. Charge order and three-site distortions in the Verwey structure of magnetite. Nature. 2012;**481**:173-176.

[54] Skumryev V, Blythe HJ, Cullen J, Coey JMD. AC susceptibility of a magnetite crystal. Journal of Magnetism and Magnetic Materials. 1999;**196-197**:515-517.

[55] Gaudisson T, Acevedo U, Nowak S, Yaacoub N, Greneche J-M, Ammar S, et al. Combining soft chemistry and spark plasma sintering to produce highly dense and finely grained soft ferrimagnetic $Y_3Fe_5O_{12}$ (YIG) ceramics. Journal of the American Ceramic Society. 2013;**96**(10):3094-3049.

[56] X. Liu WZ, S. Yang, Z. Yu, B. Gu, Du Y. Structure and magnetic properties of La^{3+}-substituted strontium hexaferrite particles prepared by sol–gel method. Physica Status Solidi A. 2002;**193**(2):314-319.

[57] Ataie A, Heshmati-Manesh S. Synthesis of ultra-fine particles of strontium hexaferrite by a modified co precipitation method. Journal of European Ceramic Society. 2001;**21**(10):1951-1955.

[58] Chen D-H, Chen Y-Y. Synthesis of strontium ferrite ultrafine particles using microemulsion processing. Journal of Colloid and Interface Science. 2001;**236**(1):41-46.

[59] Naiden EP, Itin VI, Terekhova OG. Mechanochemical modification of the phase diagrams of hexagonal oxide ferrimagnets. Technical Physics Letters. 2003;**29**(11):889-891.

[60] Kurisu S, Kubo O, editors. Ferrites. Tokyo and Kyoto, Japan: ICF6; 1992.

[61] Taguchi H, Takeishi V, Suwa K, Masuzawa K, Minachi Y. High energy ferrite magnets. Journal de Physique IV France. 1997;**7**(C1):311-312.

[62] Obara G, Yamamoto H, Tani M, Tokita M. Magnetic properties of spark plasma sintering magnets using fine powders prepared by mechanical compounding method. Journal of Magnetism and Magnetic Materials. 2002;**239**(1-3):464-467.

[63] Boda SK, Thrivikraman G, Panigrahy B, Sarma DD, Basu B. Competing roles of substrate composition, microstructure, and sustained strontium release in directing osteogenic differentiation of hMSCs. ACS Applied Materials & Interfaces. 2016;**8**:31567-31573.

[64] Ovtar S, Le Gallet S, Minier L, Millot N, Lisjak D. Control of barium ferrite decomposition during spark plasma sintering: Towards nanostructured samples with anisotropic magnetic properties. Journal of the European Ceramic Society. 2014;**34**:337-346.

[65] Bolarin AM, Sanchez De Jesus F, Cortes-Escobedo CA, Díaz-De la Torre S, Valenzuela R. Synthesis of M-type $SrFe_{12}O_{19}$ by mechanosynthesis assisted by spark plasma sintering. Journal of Alloys and Compounds. 2014;**643**(S1):S226-S230.

[66] Dehghan R, Seyyed Ebrahimi SA. Optimized nanocrystalline strontium hexaferrite prepared by applying a methane GTR process on a conventionally synthesized powder. Journal of Magnetism and Magnetic Materials. 2014;**368**:234-239.

[67] Zhao W-Y, Zhang Q-J, Tang X-F, Cheng H-B, Zhai P-C. Nanostructural M-type barium hexaferrite synthesized by spark plasma sintering method. Journal of Applied Physics. 2006;**99**(8):08E909.

[68] Saura-Muzquiz M, Granados-Miralles C, Stingaciu M, Bojesen ED, Li Q, Song J, et al. Improved performance of $SrFe_{12}O_{19}$ bulk magnets through bottom-up nanostructuring. Nanoscale. 2016;**8**:2857-2866.

[69] Stingaciu M, Topole M, McGuiness P, Christensen M. Magnetic properties of ball-milled $SrFe_{12}O_{19}$ particles consolidated by spark-plasma sintering. Nature, Scientific Reports. 2015;5:14112.

[70] Spaldin NA, Fiebig M. The renaissance of magnetoelectric multiferroics. Materials Science. 2005;**309**(5733):391-392.

[71] Jiang QH, Shen ZJ, Zhoua JP, Shia Z, Nan CW. Magnetoelectric composites of nickel ferrite and lead zirconnate titanate prepared by spark plasma sintering. Journal of the European Ceramic Society. 2007;**27**(1):279-284.

[72] Stingaciu M, Kremer RK, Lemmens P, Johnsson M. Magnetoresistivity in $CoFe_2O_4$-$BaTiO_3$ composites produced by spark plasma sintering. Journal of Applied Physics. 2011;**110**(4):044903.

[73] Liu Y, Ruan X, Zhu B, Chen S, Lu Z, Shi J, et al. $CoFe_2O_4/BaTiO_3$ Composites via spark plasma sintering with enhanced magnetoelectric coupling and excellent anisotropy. Journal of the American Ceramic Society. 2011;**94**(6):1695-1697.

[74] Ghosh D, Han H, Nino JC, Subhash G, Jones JL. Synthesis of $BaTiO_3$ 20wt%$CoFe_2O_4$ nanocomposites via spark plasma sintering. Journal of the American Ceramic Society. 2012;**95**(8):2504-2509.

[75] Etier M, Schmitz-Antoniak C, Salamon S, Trivedi H, Gao Y, Nazrabi A, et al. Magnetoelectric coupling on multiferroic cobalt ferrite–barium titanate ceramic composites with different connectivity schemes. Acta Materialia. 2015;**15**:1-9.

[76] Lopez Noda R, Acevedo Salas U, Gaudisson T, Piñar FC, Ammar S, Valenzuela R. Magnetoelectric coupling in $BaTiO_3$–$CoFe_2O_4$ nanocomposites studied by impedance spectroscopy under magnetic field. IEEE Transactions on Magnetics. 2014;**50**(11):8002304.

[77] Acevedo U, Lopez-Noda R, Breitwieser R, Calderon F, Ammar S, Valenzuela R. An impedance spectroscopy study of magnetodielectric coupling in $BaTiO_3$-$CoFe_2O_4$ nanostructured multiferroics. AIP Advances. 2017;**7**:055813.

[78] Acevedo U, Unpublished results.

[79] Loyau V, Morin V, Chaplier G, LoBue M, Mazaleyrat F. Magnetoelectric effect in layered ferrite/PZT composites. Study of the demagnetizing effect on the magnetoelectric behavior. Journal of Applied Physics. 2015;**117**(18):184102.

[80] Dai Z, Akishige Y. Electrical properties of $BiFeO_3$–$BaTiO_3$ ceramics fabricated by mechanochemical synthesis and spark plasma sintering. Materials Letters. 2012;**88**:36-39.

[81] Stingaciu M, Reuvekamp PG, Tai C-W, Kremer RK, Johnsson M. The magnetodielectric effect in $BaTiO_3$–$SrFe_{12}O_{19}$ nanocomposites. Journal of Materials Chemistry C. 2014;**2**:325-330.

Chapter 3

Phase Identification and Size Evaluation of Mechanically Alloyed Cu-Mg-Ni Powders

Celal Kursun, Musa Gogebakan,
M. Samadi Khoshkhoo and Jürgen Eckert

Additional information is available at the end of the chapter

http://dx.doi.org/10.5772/67404

Abstract

Ternary mixture of Cu, Mg, and Ni with the nominal composition of nanocrystalline $Cu_{50}Mg_{25}Ni_{25}$ (in at.%) was milled for 25 hours. Analysis of an X-ray diffraction pattern (XRD) and transmission electron microscopy (TEM) was used to characterize the chemical phases and microstructure of the final product, which is shown to consist of ternary alloy of Cu-Mg-Ni with FCC structure along with small amounts of FCC MgO and $Mg_{0.85}Cu_{0.15}$. The good agreement between the size values obtained by XRD and TEM is attributed to the formation of defect-free grains with no substructure during ball milling. Dynamic recrystallization may be a possible mechanism for the emergence of such small grains (<20 nm). The particle size distribution and morphological changes of Cu–Mg–Ni powders were also analyzed by scanning electron microscopy (SEM). According to the SEM results, the particle size of the powders decreased with increasing milling time. Lattice parameter of the Cu-Mg-Ni ternary FCC alloy formed during mechanical alloying increased with increase in milling time from 3.61 to 3.65 Å after 20 hours milling.

Keywords: nanocrystalline powders, Cu-Mg-Ni alloy, mechanical alloying, Rietveld, TEM

1. Introduction

High strength and good electrical and thermal conductivities are the fundamental requirements for materials in electrical industries [1]. Copper-based alloys are the optimal materials for such applications [2]. There are certain methods to fabricate these materials such as mechanical alloying, vapor deposition, rapid solidification, etc. In mechanical alloying, materials are obtained in powder form, which can be later compacted to desired shapes and dimensions for practical applications. The advantage of mechanical alloying is the possibility and easy production of super saturated solid solutions and meta-stable phases that is generally difficult to obtain from other techniques [3, 4]. Thus, many studies have been devoted to the production of meta-stable materials that are amorphous, nanocrystalline, and quasicrystalline by using mechanical alloying.

Mechanical and physical properties of materials strongly depend on their grains size [5, 6]. A lot of studies have been carried out to enlighten the relationship between the microstructure and mechanical and physical properties [7–10]. For this purpose, determination of grain size is of great importance. Transmission electron microscopy (TEM) investigation, that is based on direct observations and counting of the grains [11], provides grain size and grain size distribution, which is closer to the reality. However, due to the ease of X-ray diffraction technique, this method has been used extensively to determine the size of the coherent domains in nanomaterials. Size evaluation using X-ray diffraction pattern (XRD) is based on the broadening of reflections in diffraction pattern. Usually, there is a large discrepancy between the results obtained by TEM and XRD [12]. This discrepancy originates from the difference between the type of the information obtained by XRD and TEM. In fact, the size value obtained by TEM and XRD corresponds to grain size and coherently scattering of domain size. Depending on the processing route, both these values can be near or far from each other. There are several methods to obtain coherent domain size and microstrain from the X-ray diffraction patterns including Williamson-Hall, Halder-Wagner, Warren-Averbach, Debye-Scherrer, and Rietveld refinement method [13]. In the present work, the microstructure of the final powder is investigated using transmission electron microscopy (TEM), Debye-Scherrer equation, and Rietveld refinement of X-ray diffraction (XRD) patterns. The good agreement obtained between the results of XRD and TEM will be discussed in terms of the possible microstructural evolutions during mechanical alloying.

2. Experimental

Ternary Cu-Mg-Ni powder alloy with the nominal composition of $Cu_{50}Mg_{25}Ni_{25}$ was mechanically alloyed in planetary ball mills (Fritsch Pulverisette 5). The elemental powders of Cu, Mg, and Ni were accurately weighted to the desired compositions. Powders together with stainless steel milling balls were charged into a stainless steel vial (125 mL). Ball milling was performed at room temperature at the rotation speed of 300 rpm with a ball to powder mass ratio (BPR) of 10:1. The powders were mechanical alloyed up to 25 hours. After each 15 minutes of ball milling, the process was interrupted for 30 minutes in order to cool down the vials. Samples were taken at suitable milling times to follow the changes of microstructure and phases during ball milling.

Phase transformation during milling was studied using X-ray diffraction (XRD) analysis. For this purpose, a diffractometer with Co Kα radiation operating at 40 kV was employed. The crystallite size determination was performed using diffraction pattern obtained by a STOE Stadi P diffractometer (CuKα1 radiation) operating at 40 kV in transmission geometry with small instrumental broadening. The diffraction patterns were recorded using a linear position sensitive detector with the 2θ (diffraction angle) range of 30–130°. TEM investigations of the final product were performed using a Phillips CM-20 transmission electron microscope operating at 200 kV.

3. Results and discussion

Figure 1 shows the X-ray diffraction patterns of $Cu_{50}Mg_{25}Ni_{25}$ powder at different milling times using X-ray source wavelength = 0.154056 nm. The reflections corresponding to the starting materials Cu, Mg, and Ni are broadened and fade away only after 5 hours of milling as shown in **Figure 1(a)**. The elemental Cu peaks decrease more rapidly than Ni peaks, indicating faster grain size refinement in the Cu powders. Similar observation is reported by Gogebakan et al. for the $Al_{65}Cu_{20}Fe_{15}$ [14]. The main reflections of Ni and Cu become closer to each other and form a single broad reflection in the 2θ range of 45–55°. This happens due to the fact that mechanical alloying pumps the atoms of the existing elements into the lattice of each other, leading to the opposite shift in the peak positions of Ni and Cu. As a result, initial crystals are also heavily strained that leads to the broadening of the reflections. On the other hand, grain refinement occurs during ball milling that causes the additional broadening of the reflections. A rough modeling of the XRD pattern using Rietveld analysis for 5 hours of milling showed that the observed broad peak for $Cu_{50}Mg_{25}Ni_{25}$ can be attributed to the (111) reflection of an FCC structure (in this case Cu-based solid solution) with coherently scattering domain.

On further milling, the broad reflection became sharper and the other reflections of lower intensity appeared. As shown in **Figure 1(b)**, finally after 25 hours of milling, the structure is composed of a Cu-based FCC structure along with a minor amount of MgO and $Mg_{0.85}Cu_{0.15}$ intermetallic. They are marked by symbols in **Figure 1(b)**.

X-ray diffraction pattern of the final powder (milled for 25 hours) was modeled using Rietveld refinement. A good agreement was obtained between the experimental and calculated patterns as shown in **Figure 2**. Only the effects of size and strain were considered during the refinement with no special corrections for the effect of planar lattice defects as well as strain anisotropy. The size values obtained for the Cu-based FCC solid solution, $Mg_{0.85}Cu_{0.15}$ intermetallic, and MgO are 10, 10, and 15 nm, respectively.

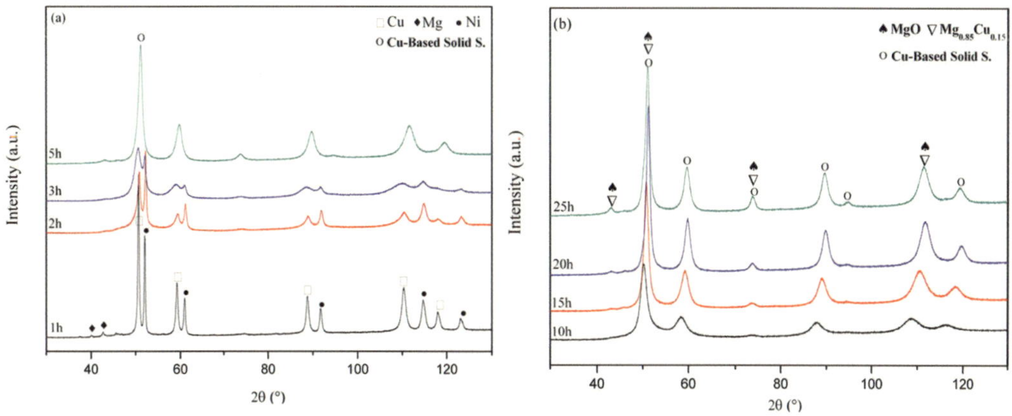

Figure 1. XRD patterns of $Cu_{50}Mg_{25}Ni_{25}$ alloy after different milling time; (a) 1–5 hours, (b) 10–25 hours.

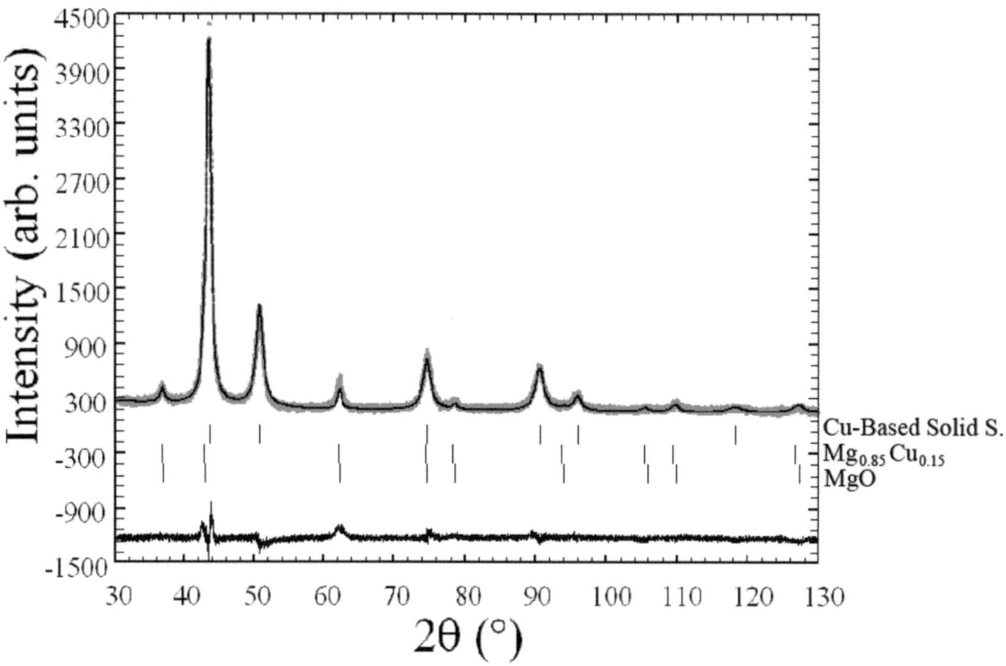

Figure 2. Rietveld refinement for XRD patterns of $Cu_{50}Mg_{25}Ni_{25}$ powders after 25 hours of milling time.

The crystallite size evolution of the $Cu_{50}Mg_{25}Ni_{25}$ powder alloy as a function of milling time is presented in **Figure 3**. The variation of crystallite size of the powder alloy was estimated by broadening of XRD peaks using Debye-Scherrer equation [15]

$$D = \frac{0.9\lambda}{B\cos\theta} \tag{1}$$

where D is the average crystallite size, λ is the wave length of using X-ray, B is the full width (in radians) at half maximum intensity, and θ is the diffraction Bragg angle.

As seen in **Figure 3**, the crystallite size of the ternary FCC Cu-Mg-Ni alloy is found to decrease initially with increase in milling time approaching near saturation after 10 hours of milling. It was calculated to be about 30, 18.5, 19, and 17 nm after 5, 10, 15, and 20 hours milling, respectively, and reached a steady-state value of about 17 nm after 25 hours of milling.

The morphological changes of the $Cu_{50}Mg_{25}Ni_{25}$ alloy during mechanical alloying are evident from SEM micrographs, which are shown in inset in **Figure 3**. From SEM micrographs, it can be seen clearly that the unmilled powders have different shapes and particle sizes. The Cu and Mg powder particles are of irregular shapes with an average size in the range of 50–150 μm. The Ni powder is a spherical morphology with size in the range of 20–60 μm. After 5 hours of milling, the powder particles of Cu–Mg–Ni alloy became nearly spherical shaped with an average size in the range of 10–30 μm. On further milling (10 hours milling), the homogeneity of the $Cu_{50}Mg_{25}Ni_{25}$ alloy increased and its particle size decreased up to 5 μm. For the higher milling time up to 25 hours, the formation of submicrometer particles was observed. Therefore, the average particle size was obviously determined below 5 μm.

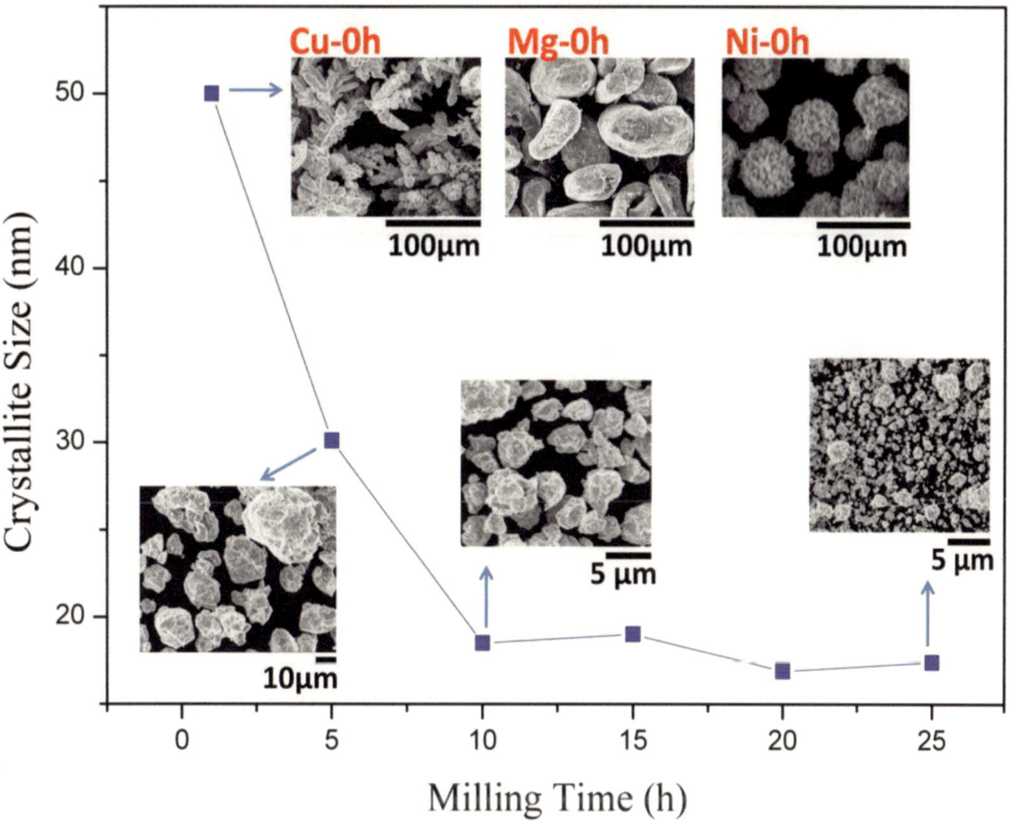

Figure 3. Crystal size of mechanically alloyed $Cu_{50}Mg_{25}Ni_{25}$ powders as a function of milling time and the insets: SEM micrographs of the powder particles after 0, 5, 10, and 25 hours of milling times.

The sample after 25 hours of milling was also investigated using transmission electron microscopy (TEM). Typical dark-field (DF) images are shown in **Figure 4(a** and **b)**. According to the TEM micrographs, the structure consists of nanocrystalline grains with the size of around 20 nm. **Figure 4(c)** illustrates the corresponding grain size distribution obtained using several DF images in order to have sufficient statistics (more than 240 grains). The obtained grain size distribution was fitted to the log-normal distribution function to calculate the mean average grain size. With the median m and the variance σ of the log-normal distribution function, the average grain sizes can be obtained according to the Eq. (2):

$$\langle x_j \rangle = m \exp(k\sigma^2) \quad (2)$$

With the k values of 0.5, 2.5, and 3.5 will give the arithmetic, area-weighted, and volume-weighted average grain size, respectively [16]. In order to compare with the results obtained by Rietveld refinement of XRD patterns and Debye-Scherrer equation, volume-weighed grain size was calculated to be 20 nm. This is in a good agreement with the values of 10 and 17.4 nm obtained by Rietveld refinement of XRD patterns and Debye-Scherrer equation, respectively. These results agree with a previous study which reported the production of similar compositions of Cu-Mg-Ni

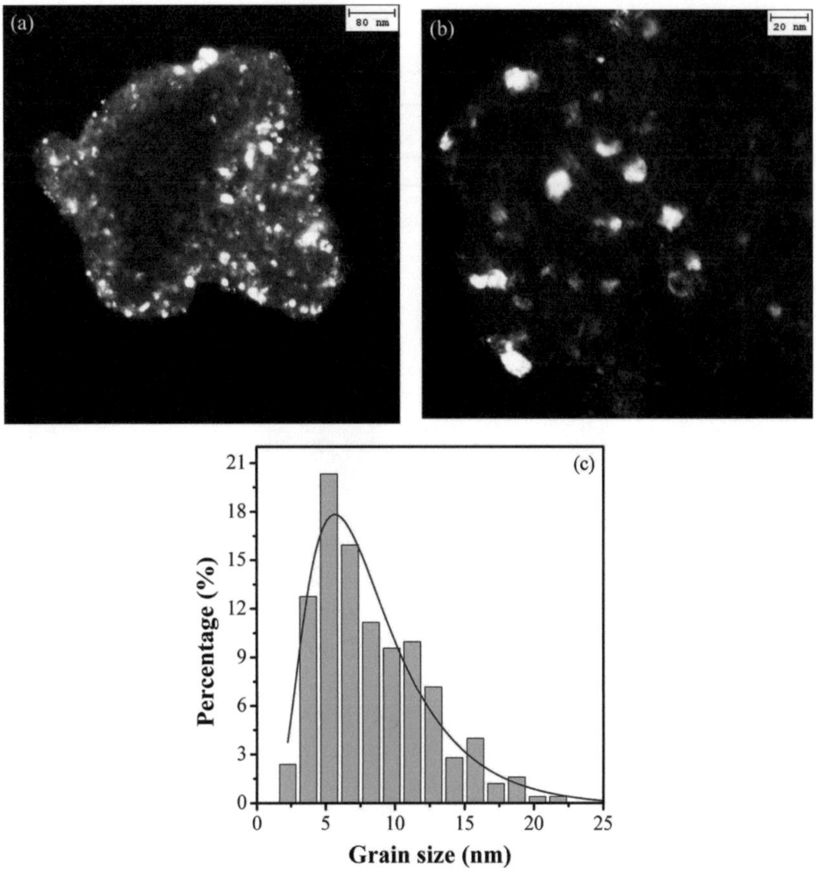

Figure 4. Dark field TEM micrographs of $Cu_{50}Mg_{25}Ni_{25}$ powders milled for (a), (b) 25 hours, and (c) the corresponding grain size distribution with the solid line as the fit to the log-normal distribution.

powders by the mechanical alloying, although the starting compositions were different being $Cu_{50}Mg_{30}Ni_{20}$ and $Cu_{50}Mg_{45}Ni_{5}$ [17].

Size calculations by TEM and XRD have been reported in many different systems, many often showing a large discrepancy between the size values obtained by TEM and XRD. Accordingly, the large discrepancy is due to the hierarchy of the deformed structure, details of which can be found elsewhere [16, 18]. This originates from the type of the information obtained by XRD and TEM, which are coherently scattering domain size (the smallest unfaulted peace of a crystal [19]) and grain size, respectively. Another reason giving rise to the discrepancy is the miscalculation of the coherent domain size by XRD that is due to the elastic anisotropy of the investigated materials [19]. The consequence of the presence of elastic anisotropy is that the strain broadening is not anymore a monotonous function of diffraction angle (2θ). New models have been developed in order to consider the effect of elastic anisotropy with the assumption that most of the strain originates from the presence of dislocations and the introduction of the concept of dislocation contrast factors [20, 21]. The effect of planar faults is also included in these models [20, 21]. The size values by XRD in this work are obtained without any special considerations about elastic anisotropy or the presence of planar faults. However, there is a good agreement between the two values. A possible explanation is that the FCC solid solution formed during milling does not have a considerable substructure inside grains. According to diffraction patterns (**Figure 1(b)**), the reflections related to the FCC solid solution do not exist before 10 hours of ball milling. They emerge after 10 hours and their intensity increases by further milling. This suggests that they may form by some dynamic processes like dynamic recrystallization of defect-free small grains within the heavily strained matrix. The observation that the reflections do not exist before 10 hours of milling and amount of that increases by further milling, strengthens the idea of recrystallization. As a result, the structure is composed of nanosized grains with no substructure. It is also clear from the TEM micrographs of the final sample. The uniform diffraction of the electron beam by grains implies that they are substructure-free. The two size values obtained by XRD and TEM are in a good agreement because XRD gives the size of the substructure and that the grains do not contain any substructure.

Figure 5 shows lattice parameters of Cu-based FCC solid solution phase for $Cu_{50}Mg_{25}Ni_{25}$ alloy that had been mechanically alloyed for various milling times. As it is seen in **Figure 5**, the lattice parameters increase with increasing milling time. The increase of lattice parameter with milling time may be attributed to the effect of dissolving of bigger Mg atoms into Cu. It can be seen that the variation is sharp at the initial stage, however, after 15 hours milling it reaches a constant value. This indicates that the dissolution of Mg into Cu is fully completed. On the other hand, the effect of Ni to lattice parameter is very low because the atomic radii of Cu (0.128 nm) and Ni (0.125 nm) are close to each other. This provides a likely explanation of why the lattice parameter

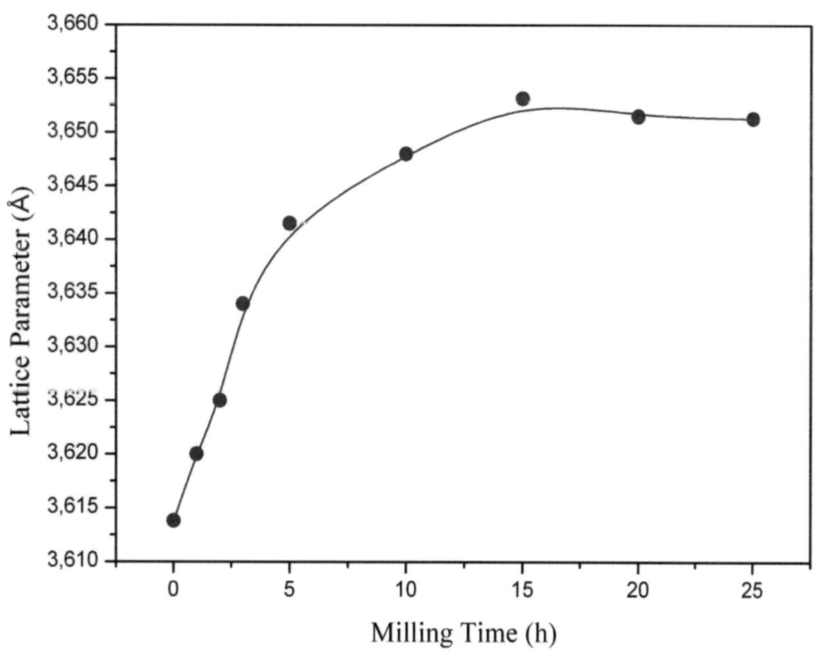

Figure 5. Lattice parameters of Cu-based FCC solid solution phase as a function of milling time. The line through the data points is a visual guide.

of Cu-based FCC solid solution phase increases from 3.61 to 3.65 Å during different stages of milling from 0 to 25 hours. For comparison, the lattice constant of MgO phase is 4.211 Å.

4. Conclusions

In this study, the nature of the phases formed and their particle sizes have been reported when the starting ternary mixture of Cu:Mg:Ni in the ratio of 50:25:25 is ball-milled for 25 hours. After 25 hours milling time, nanostructured Cu-based solid solution containing Cu, Mg, and Ni with FCC structure along with smaller quantities of FCC $Mg_{0.85}Cu_{0.15}$ and FCC MgO phases is formed. The crystallite size of this alloy was calculated by Debye-Scherrer and Rietveld Refinement methods using XRD data. In order to confirm the crystallite size obtained by XRD, the microstructure of the final powder was also monitored by TEM. The size value obtained by the TEM was determined to be 20 nm, which is in a good agreement with values determined from the analysis of the XRD pattern. This agreement is attributed to the formation of defect-free grains with no substructure during ball milling. It may be a possible mechanism of dynamic recrystallization for the emergence of such small grains (<20 nm). The morphology and particle size distribution of the $Cu_{50}Mg_{25}Ni_{25}$ alloy have been changed during mechanical alloying. The elemental powders, which have different shapes in the initial stage, became nearly rounded and the homogeneity increased with increasing milling time. The average particle size of the final product was determined to be below 5 μm. The remaining unresolved issues are the exact atomic concentrations of Cu, Mg, and Ni in the ternary alloy formed in this process here and in earlier studies reported in Ref. [17] and the nature of the phase diagram of the Cu, Mg, Ni system. These issues will be investigated in future studies.

Acknowledgements

The authors thank Kahramanmaras Sutcu Imam University for financial support of the research program (Project No:2011/3-21). One of the authors (C. Kursun) would like to thank the Council of Higher Education (YÖK) for graduate research support.

Author details

Celal Kursun[1]*, Musa Gogebakan[1], M. Samadi Khoshkhoo[2] and Jürgen Eckert[3]

*Address all correspondence to: celalkursun@ksu.edu.tr

1 Department of Physics, Faculty of Art and Sciences, Kahramanmaras Sutcu Imam University, Kahramanmaras, Turkey

2 IFW Dresden, Institute for Complex Materials, Dresden, Germany

3 Department of Materials Physics, Erich Schmid Institute of Materials Science, Austrian Academy of Sciences (ÖAW), Leoben, Austria

References

[1] M. Azimi, G.H. Akbari, Characterization of nano-structured Cu–6 wt.% Zr alloy produced by mechanical alloying and annealing methods, J Alloy Compd 555 (2013) 112-116.

[2] D. Zhao, Q.M. Dong, P. Liu, B.X. Kang, J.L. Huang, Z.H. Jin, Aging behavior of Cu–Ni–Si alloy, Mat Sci Eng A-Struct 361 (2003) 93-99.

[3] C. Suryanarayana, Mechanical alloying and milling, Prog Mater Sci 46 (2001) 1-184.

[4] C. Kursun, M. Gogebakan, Characterization of nanostructured Mg–Cu–Ni powders prepared by mechanical alloying, J Alloy Compd 619 (2015) 138-144.

[5] M.A. Meyers, A. Mishra, D.J. Benson, Mechanical properties of nanocrystalline materials, Prog Mater Sci 51 (2006) 427-556.

[6] T. Varol and A. Canakci, Effect of particle size and ratio of B_4C reinforcement on properties and morphology of nanocrystalline Al2024-B_4C composite powders, Powder Technol 246 (2013) 462-472.

[7] K. Shengzhong, F. Lui, D. Yutian, X. Guangji, D. Zongfu, L. Peiqing, Synthesis and magnetic properties of Cu-based amorphous alloys made by mechanical alloying, Intermetallics 12 (2004) 1115-1118.

[8] G. Wang, S. Fang, X. Xiao, Q. Hua, J. Gu, Y. Dong, Microstructure and properties of $Zr_{65}Al_{10}Ni_{10}Cu_{15}$ amorphous plates rolled in the super-cooled liquid region, Mater Sci and Engg A 373 (2004) 217-220.

[9] M. Gogebakan, The effect of Si addition on crystallization behaviour of amorphous Al-Y-Ni alloy, J Mater Eng Perform 13 (2004) 504-508.

[10] S. Deledda, J. Eckert, L. Schultz, Mechanical alloyed Zr-Cu-Al-Ni-C glassy powders, Mat Sci Eng A-Struct 375-377 (2004) 804-808.

[11] J. Guerrero-Paz, D. Jaramillo-Vigueras, Comparison of grain size distributions obtained by XRD and TEM in milled FCC powders, Nanostrcut Mater 11 (1999) 1195-1204.

[12] Y. Zhong, D.H. Ping, X.Y. Song, F.X. Yin, Determination of grain size by XRD profile analysis and TEM counting in nano-structured Cu, J Alloy Compd 476 (2009) 113-117.

[13] F. Hadef, A. Otomani, A. Djekoun, J.M. Grenèche, Structural and microstructural study of nanostructured $Fe_{50}Al_{40}Ni_{10}$ powders produced by mechanical alloying, Mater Charact 62 (2011) 751-759.

[14] M. Gogebakan, B. Avar, Quasicrystalline phase formation during heat treatment in mechanically alloyed $Al_{65}Cu_{20}Fe_{15}$ alloy, Mater Sci Tech Ser 26 (2010) 920-924.

[15] C. Suryanarayana, M. Grant Norton, X-ray Diffraction A Practical Approach, Plenum Press, New York (1998), p. 207.

[16] T. Ungár, G. Tichy, J. Gubicza, R.J. Hellmig, Correlation between subgrains and coherently scattering domains, Powder Diffr 20 (2005) 366-375.

[17] M. Gogebakan, C. Kursun, J. Eckert, Formation of new Cu-based nanocrystalline powders by mechanical alloying technique, Powder Technol 247 (2013) 172-177.

[18] M. Samadi Khoshkhoo, S. Scudino, J. Thomas, K.B. Surreddi, J. Eckert, Grain and crystallite size evolution of cryomilled pure copper, J Alloy Compd 509 (2011) 343–S347.

[19] E. Schafler, M. Zehetbauer, Characterization of nanostructured materials by X-ray line profile analysis, Rev Adv Mater Sci 10 (2005) 28-33.

[20] T. Ungár, A. Borbély, The effect of dislocation contrast on X-ray line broadening: A new approach to line profile analysis, Appl Phys Lett 69 (1996) 3173-3175.

[21] T. Ungár, Á. Révész, A. Borbély, Dislocations and grain size in electrodeposited nanocrystalline Ni determined by the modified Williamson-Hall and Warren-Averbach procedures, J. Appl. Cryst 31 (1998) 554-558.

Chapter 4

Wear Mechanism and Failure of Carbide Cutting Tools with Nanostructured Multilayered Composite Coatings

Alexey A. Vereschaka, Sergey N. Grigoriev,
Nikolay N. Sitnikov and Gaik V. Oganyan

Additional information is available at the end of the chapter

http://dx.doi.org/10.5772/68008

Abstract

The aim of this work is to study physical and chemical properties of nanostructured multi-layered composite coating based on three-layered architecture, deposited to a carbide substrate, as well as to study the mechanism of wear and failure of coated carbide tools under the conditions of stationary cutting. The coating were obtained by the method of filtered cathodic vacuum arc deposition (FCVAD). Here, the microstructure of coating as well as its hardness, strength of the adhesive bond to the substrate, chemical composition and phase composition were investigated on a transverse cross-section of experimental samples. The studies of cutting properties of the carbide inserts with developed coatings was conducted on a lathe in longitudinal turning of steel C45 (HB 200). The analysis of mechanisms of wear and failure of coated tool was carried out, including the processes of diffusion and oxidation in the surface layers of the coated substrate. Tools with harder and less ductile coatings showed less steady kinetics of wear, characterized by sharp intensification of wear and failure in transition from "steady" to drastic wear, i.e., at the end of the tool life. The X-ray microanalysis showed a considerable increase in oxygen content in the transverse cracks in the coating.

Keywords: nanostructured multilayer composite coatings, tool life, carbide cutting tool, crack formation, filtered cathodic vacuum arc deposition

1. Introduction

The efficiency of the cutting tool is a function of the complex processes of contact interaction between the tool material and the material being machined. This efficiency is determined by various factors, including: (i) the crystal-chemical, thermal-physical, and mechanical properties of the material being machined and the tool material. Depending on the conditions of their contact, for example, dry cutting, cutting with CF (define CF) and type of machining operation, the level of thermomechanical loading on the cutting part of the tool may be influenced; and (ii) the geometrical dimensions of the cutting part of the tool, machining conditions, and the kinematics of motion of interacting tool and workpiece surfaces.

In accordance with the above, during the cutting process, the contact areas of the coated tool are exposed to a combination of factors that trigger their wear and fracture. Such factors usually include (i) an abrasive effect between the material being machined and the coated tool material; (ii) adhesive and adhesive-fatigue processes at the boundaries of the contact between the coated tool material and the material being machined, resulting in "ultimate fatigue" of local volumes, the tool material coating, the formation of cross cracks from fatigue, delamination, and finally the removal of fractured fragments of coating and damaged volumes of tool material (substrate) with trailing chips; (iii) macro- and microchips, as well as brittle fracture of local volumes of the tool material coating, resulting from exposure to pulsed shock loads; (iv) fracture of macrovolumes of the tool material and the material being machined as a result of the combined effect of alternating mechanical and thermal stresses; (v) corrosion and oxidation processes; and (vi) diffusion interaction between the tool material and the material being machined.

Depending on the cutting conditions, the above factors may become either dominant or play a secondary role. Until recently, many of the factors with strong influences on the wear and failure of coated tools were not well known. Therefore, a brief analysis of the publications devoted to research of mechanisms of wear/fatigue processes of cutting tools with complex composite coatings, formed with the use of the vacuum-arc technologies (arc-PVD), is presented below.

Harry et al. [1] studied the mechanism of cracking in tungsten-carbon–based multilayered coatings. The analysis of cracking probability showed a significant effect of the number and thickness of the elementary layers on the fracture resistance of the multilayered coatings. Mo et al. [2] and Birol [3] presented studies of coatings based on CrN, AlCrN, and AlTiN. Impact wear tests were performed to investigate the impact resistance of the coatings. There was no visible crack formation for all three coatings during the impact tests. The AlCrN coating exhibited the best impact wear performance. Nohava et al. [4] studied the properties of AlTiNAlCrN, AlCrON, and α-$(Al,Cr)_2O_3$ coatings. The main wear mechanisms at high temperature were oxidative attack accompanied by gradual material removal. Aihua et al. [5] studied the properties of the coatings TiN, TiAlN, AlTiN, and CrAlN. The wear mechanism of TiAlN was a combination of abrasive wear, oxidation, partial fracture, and microgroove formation. The AlTiN coating was worn by plowing and mechanical-dominated wear, and the damage was caused via a brittle failure mechanism. CrAlN coating presented the best properties of antispalling and antiadhesion. Antonov et al. [6] focused on the properties of gradient and multilayer coatings (AlTiN-Si_3N_4 and also TiN, TiCN, TiAlN, AlTiN). Erosive, abrasive, and impact wear tests were conducted. Henry et al. [7] studied the properties of six titanium aluminum nitrides $Ti_{1-x}Al_xN$ ($0 \leq x \leq 1$). Ti-rich and Al-rich films present two different wear mechanisms related to their toughness values.

http://dx.doi.org/10.5772/68008

To study the properties of coatings, a number of the papers used the method of nanoimpact testing. In particular, in the paper prepared by Beake et al. [8], this method was used to study the coatings $Ti_{1-x}Al_xN$ (x = 0.5 and 0.67). Skordaris et al. [9] studied multilayer TiAlN coatings with an Al/Ti ratio of 54/46. The coatings consisted of one, two, or four layers. Beake et al. [10] studied wear performance of end mill tools with AlCrN monolayer and AlCrN–TiAlN bilayer coatings. The impact fatigue fracture resistance of the AlCrN monolayer coating was lower than AlTiN and AlCrN–TiAlN bilayers (by nanoimpact test). In Ning et al. [11], a number of nanomultilayered TiAlCrN/M_{ex}N coatings (where M_{ex} are transition metals such as Nb, Ta, Cr, W) and monolayer TiAlCrN coatings were tested during dry, high-speed end milling of hardened AISI H13 (HRC 55–57). The dominant wear mechanism for the worn coating zones was abrasion wear. The studies presented by Fox-Rabinovich et al. [12] were focused on the properties of the coatings on the basis of systems TiAlCrN, TiAlCrN/TaN, TiAlCrN/CrN, TiAlCrN/WN, and TiAlCrN/NbN nanomultilayered coatings. In another study, Fox-Rabinovich et al. [13] studied the properties of AlTiN and TiAlCrN PVD coatings. According to the authors [12], in many cases, the hardness of coatings can be a marginal property where the major properties are the plasticity and impact fatigue fracture resistance; a surface with these characteristics is able to dissipate energy by means of plastic deformation and thus surface damage and wear rate are reduced.

The wear propagation of a TiAlN coating during the impact test at temperatures up to 400°C was monitored by Erkens et al. [14] in terms of the coating fracture ratio (FR) versus the applied impact force. Beake and Fox-Rabinovich [15] applied high-temperature nanomechanical testing to the study of PVD coatings (AlTiCrSiYN, AlTiCrN, AlTiN, AlCrN, among others), to help explain why certain coating compositions work well in some applications, but not others. The hard-to-cut materials were being machined during high-speed machining tests. A recently developed multilayer coating, AlTiCrSiYN-AlTiCrN, with a combination of mechanical properties and microstructure can minimize crack formation and dissipate energy by crack deflection along interfaces. The studies were also focused on different mechanical properties of coatings such as (Ti, Cr, Al, Si)N [16], (Al,Cr,Ta,Ti,Zr)N [17], (Al,Si,Ti)N [18], Ti-TiN-(TiCrAl)N [19, 20], Zr-(Zr,Cr)N-CrN and Ti-TiN-(Ti,Cr,Al)N [21], and Ti-(AlCr)N-(TiAl)N, Ti-(AlCr)N-(TiCrAl)N, and Zr-(AlCr)N-(ZrCrAl)N [22].

The review prepared by Bouzakis et al. [23] presents a detailed overview of the available methods to control basic properties of wear-resistant coatings, including control over crack resistance. It should be noted that cracking of the coatings has a decisive influence on failure mechanism and performance of the cutting tool. Meanwhile, only a small number of papers are focused on the study of cracking in the coatings and their influence on performance of the tool. In particular, Tabakov et al. [24, 25] discuss the cracking mechanisms with regard to monolayered macrosize coatings on the basis of systems TiN, TiCN, (Ti,Zr)N, and (Ti,Zr)CN. It was found that the coatings of complex composition of (Ti,Zr)N and (Ti,Zr)CN types better resisted intense cracking in the coating material. Tabakov et al. also studied the multilayered coatings with macrosize structure, in particular, on the basis of systems TiCN-(Ti,Zr)N-TiN, TiN-(Ti,Zr)N-TiN, TiCN-(Ti,Al)N-TiN, and TiCN-(Ti,Mo)N-TiN [26]. The conducted studies have revealed that the introduction of zirconium nitride in the coating composition sufficiently reduced the tendency of cracking. These data were obtained from the study of wear mechanisms of coated tool in milling of different materials under different conditions of cutting. The problems of cracking and brittle fracture of coatings Ti-TiN-(TiCrAl)N, Zr-(Zr,Cr)N-CrN and Ti-TiN-(Ti,Cr,Al)N; Ti-(AlCr)N-(TiAl)N, Ti-(AlCr)N-(TiCrAl)N, and Zr-(AlCr)N-(ZrCrAl)N were also discussed elsewhere [27–29].

The actual area of the surfaces of the material being machined and the tool material (including coated tool) in contact during the cutting process is less than their nominal contact area, which is predetermined by contact of vertical deviations of roughness. In the areas of actual contact, local specific pressure reaches a value at which plastic flow of metal occurs and thus the adhesion between the contacting materials dramatically increases. The initiation and failure of the adhesive bond bridging occur at high frequency of up to several thousand of failure per minute [30]. It is found out that the sizes of some adhesive spots vary from several micrometers up to several hundredths of a millimeter, and the actual contact area can reach 10–60% of the nominal contact area, while on a meter of cutting path, each contact point is exposed to shear stresses. Thus, this process determines the alternating loading of contact areas on front and flank faces of the tool and adhesive-fatigue nature of their macro- and microfracture (wear).

Tool materials are anisotropic and have various defects (inhomogeneity of structure, presence of pores, cracks, uneven distribution of residual stresses, inhomogeneity of chemical composition, etc.). Consequently, during the cutting process, along with the cut and separation of the material being machined, separation of fragments of the tool material particles also occurs. This separation is intensified as a result of the fatigue processes mentioned above and is defined as adhesive and adhesive-fatigue wear.

The most effective way to reduce these types of tool wear is to improve physical and mechanical properties of the tool material and, above all, its hardness by deposition of wear-resistant coatings. Furthermore, coatings with less physical and chemical affinities with respect to the materials being machined reduce the intensity of adhesion processes and significantly increase the wear of tool contact areas.

In an explanation of wear by the theory of adhesive-fatigue process, the separation of fragments from harder tool metal occurs, like in adhesive wear, because of presence of cracks. Cracks are formed because of the influence of two key factors: (i) repetitive mechanical stresses of a cyclical nature; and (ii) thermal stresses in the tool material characterized by different value depending on the distance from the surface of the tool contact areas. At periodic strain and compression of the upper layers of the tool material caused by mechanical or thermal influence, fatigue microcracks appear; further development then results in the growth and coalescence of microcracks, which causes separation of the fragment of the tool material and its subsequent failure.

The foregoing assumptions related to mechanisms of tool wear can serve as the basis for the formation of following requirements for coatings for cutting tools: (i) coating for a cutting tool should be structured in such a way that they form residual compressive stresses, which may act as a barrier for crack growth; (ii) coatings should have a multilayered structure so that their boundaries serve as additional barriers to crack growth; and (iii) the processes of formation of the coatings should contribute to formation of nanoscale grains and thicknesses of

http://dx.doi.org/10.5772/68008

coating sublayers, creating an optimum combination of high hardness and thermal resistance at sufficiently high fracture toughness.

2. Materials and methods

For the deposition of nanostructured carbide insert (CI) Scanning electron microscope (SEM) multilayered composite coatings (NMCC), a vacuum-arc VIT-2 unit was used, which was designed for the synthesis of coatings on substrates of various tool materials. The unit was equipped with an arc evaporator with a filtration of vapor-ion flow, which in this study was named as filtered cathodic vacuum arc deposition (FCVAD) [16–19] and was used for deposition of coatings on tool in order to significantly reduce the formation of the droplet phase during the formation of coating. The use of FCVAD process does not cause structural changes in carbide. Also it provides (i) high adhesive strength of the coating in relation to the carbide substrate; (ii) control of the level of the "healing" of energy impact on surface defects in carbide in the form of microcracks and micropores and formation of favorable residual compressive stresses in the surface layers of the carbide material; and (iii) formation of the nanoscale structure of the deposited coating layers (grain size, sublayer thickness) with high density due to the energy supplied to the deposited condensate and transformation of the kinetic energy of the bombarding ions into thermal energy in local surface volumes of carbide material at an extremely high rate of about 10^{14} K s^{-1}.

When choosing the composition of NMCC layers, forming the coating of three-layered architecture, the Hume-Rothery rule was used, which states that the difference in atomic dimensions in contacting compounds should not exceed 20% [31]. The parameters used at each stage of the deposition process of NMCC are shown in **Table 1**. An uncoated carbide tool and a carbide tool with "reference" coating TiN, deposited through the use of standard vacuum-arc technology of arc-PVD, were used as objects for comparative studies of tool life.

For microstructural studies of samples of carbide with coatings, a raster electron microscope FEI Quanta 600 FEG was used. The studies of chemical composition were conducted with the

Process	p_N (Pa)	U (V)	I_{Al} (A)	I_{ZrNb} (A)	I_{Ti} (A)	I_{Cr} (A)
Pumping and heating of vacuum chamber	0.06	+20	120	80	65	75
Heating and cleaning of products with gaseous plasma	2.0	100 DC/900 AC f = 10 kHz, 2:1	80	–	–	–
Deposition of coating	0.36	−800 DC	160	75	55	70
Cooling of products	0.06	–	–	–	–	–

I_{Ti} = current of titanium cathode, I_{Al} = current of aluminum cathode, I_{ZrNb} = current of zirconium-niobium cathode, I_{Cr} = current of chromium cathode, p_N = gas pressure in chamber, U = voltage on substrate.

Table 1. Parameters of stages of the technological process of deposition of NMCC.

use of the same raster electron microscope. To perform X-ray microanalysis, the study used characteristic X-ray emissions resulting from electron bombardment of a sample.

The hardness (HV) of coatings was determined by measuring the indentation at low loads according to the method of Oliver and Pharr [32], which was carried out on micro-indentometer Micro-Hardness Tester (CSM Instruments) at a fixed load of 300 mN. The penetration depth of the indenter was monitored so that it did not exceed 10–20% of the coating thickness to limit the influence of the substrate.

The adhesion characteristics were studied on a Nanovea scratch-tester, which represents a diamond cone with apex angle of 120° and radius of top curvature of 100 μm. The tests were carried out with the load linearly increasing from 0.05 to 40 N. Crack length was 5 mm. Each sample was subjected to three trials. The obtained curves were used to determine two parameters: the first critical load, L_{C1}, at which first cracks appeared in coating and the second critical load, L_{C2}, which caused the total failure of coating.

The studies of cutting properties of the tool made of different grades of carbide with developed NMCC were conducted on a lathe CU 500 MRD in longitudinal turning of steel C45 (HB 200). The study used cutters with mechanical fastening of inserts made of carbide (WC + 15% TiC + 6% Co) with square shape (SNUN ISO 1832:2012) and with the following figures of the geometric parameters of the cutting part: $\gamma = -8°$; $\alpha = 6°$; $K = 45°$; $\lambda = 0$; $R = 0.8$ mm. The study was carried out at the following cutting modes: $f = 0.2$ mm/rev; $a_p = 1.0$ mm; and $v_c = 250$ m/min. Flank wear-land values (VB_c) were measured with toolmaker's microscope MBS-10 as the arithmetic mean of four to five tests and a value of $VB_c = 0.45$–0.5 mm was taken as failure criteria.

3. Results and discussion

3.1. Determination of basic properties of NMCC under study

The study was focused on the NMCC containing nitrides of Ti, Al, Cr, Zr, and Nb in its composition. For the detailed studies of various properties, NMCC were selected based on the following conditions: (i) if earlier studies showed a significant increase in cutting properties and reliability of the tool [19–22, 27–29] and (ii) if the thermodynamic criterion ΔG_r (Gibbs free energy change per mole of reaction) favored the formation of the NMCC.

In order to meet the research tasks, NMCC of various compositions were selected to meet the above conditions. NMCC Cr-CrN-(TiCrAl)N, Zr-ZrN–(NbZrTiAl)N, Zr-ZrN-(ZrCrAl)N, and Ti-TiN-(ZrNbTi) were deposited using the FCVAD technology. The thicknesses of the coatings used in the studies were 2.44–3.55 μm, and a wide range of thicknesses were selected to study the effect of coating thickness on the nature of cracking. The basic properties of the NMCC under study are presented in **Table 2**. Curves obtained by mathematical processing of the experimental data are shown in **Figure 1**.

It was found out that for longitudinal turning of steel under preset cutting conditions, the carbide tools with the NMCC under study had fairly close values of tool life of about 31–37 min,

#	Composition of NMCC	Tool life T_c (min) VB = 0.4 mm	Sublayer thickness (nm)	Total thickness (μm)	Adhesion, L_{C2}, N	Hardness, HV, GPa
1	Uncoated	8	–	–	–	18
2	TiN	18	–	2.85	31	30
3	Cr-CrN-(TiCrAl)N	31	30–45	2.75	>40	38
4	Ti-TiN-(NbZrTiAl)N	35	45–60	3.30	>40	34
5	Zr-ZrN-(ZrCrAl)N	37	15–45	2.44	39	36
6	Ti-TiN-(ZrNbTi)	38	55–75	3.55	32	35

Table 2. The basic properties of the NMCC.

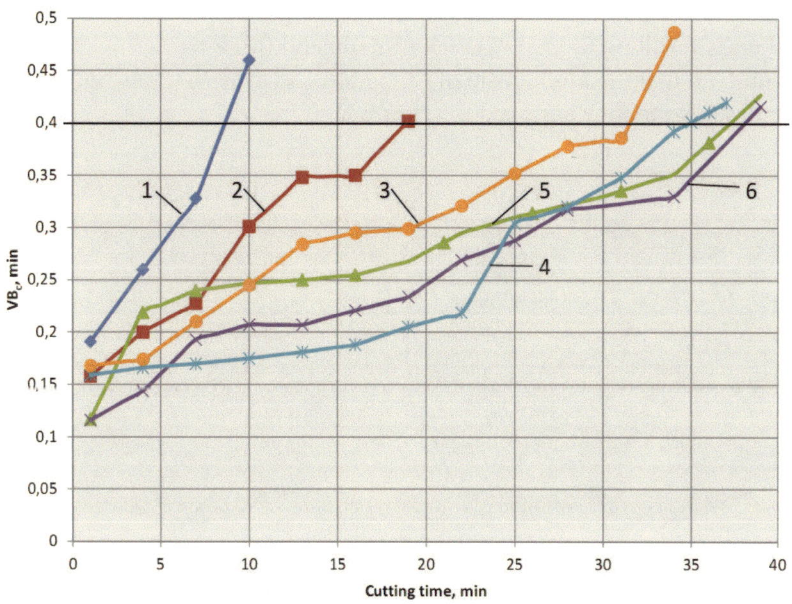

Figure 1. Dependence of wear VB on cutting time at dry turning of steel C45 at a_p = 1.0 mm; f = 0.2 mm/rev; v_c = 250 m min^{-1}. 1—Uncoated; 2—TiN; 3—Cr-CrN-(TiCrAl)N; 4—Zr-ZrN–(NbZrTiAl)N; 5—Zr-ZrN-(ZrCrAl)N; 6—Ti-TiN-(ZrNbTi).

while the nature of wear of the carbide tools with the different NMCCs under study had significant differences. In particular, NMCC Zr-ZrN-(ZrCrAl)N provided the greatest increase in tool life, reaching 37 min at stable kinetics of wear on the stages of running-in, steady, and catastrophic wear (see **Figure 1**). Almost the same stable kinetics of wear was shown by tools with NMCC Zr-ZrN–(NbZrTiAl)N, which had slightly shorter tool life of 35 min (see **Figure 1**). The tools with NMCC Ti-TiN-(ZrNbTi) also showed a long tool life of 38 min. Meanwhile, the tests showed less uniform kinetics of tool wear, which had the following specifics. It was shown that, if during 25 min of cutting (running-in and steady stages of wear) the tool wear rate of tool with NMCC Zr-ZrN–(NbZrTiAl) was lower than for the tool with NMCC Zr-ZrN-(ZrCrAl)N and Ti-TiN-(ZrNbTi), then after 25 min of cutting the wear of the tool with NMCC Zr-ZrN–(NbZrTiAl)N was sharply intensified (see **Figure 1**). The tool with NMCC Cr-CrN-(TiCrAl)N had the shortest tool life among the group of tools under study, for which 31 min of fairly uniform wear was followed by obvious intensification of wear of the tool.

3.2. Study of mechanism of adhesive-fatigue wear of carbide tool with developed NMCC

During the research of the mechanism of adhesion-fatigue wear of the carbide tool with developed NMCC, special attention was paid to the study of cracking. It was found out that the cracks under study usually had a width of several nm to 1 μm, so it is not possible to visually detect such cracks on the working surfaces of the tool by standard optical instruments. Therefore, such microcracks were detected and studied on cross-cut sections with the use of a SEM. Three main types of formed cracks were identified: (i) cross cracks; (ii) cross cracks combined with longitudinal cracks (delamination), including transformation of cross cracks into longitudinal ones, resulting in stop of crack growth; and (iii) longitudinal cracks (delamination). It is found out that the mechanism of formation and growth (branching) of cross cracks usually includes three to four basic stages (**Figure 2**):

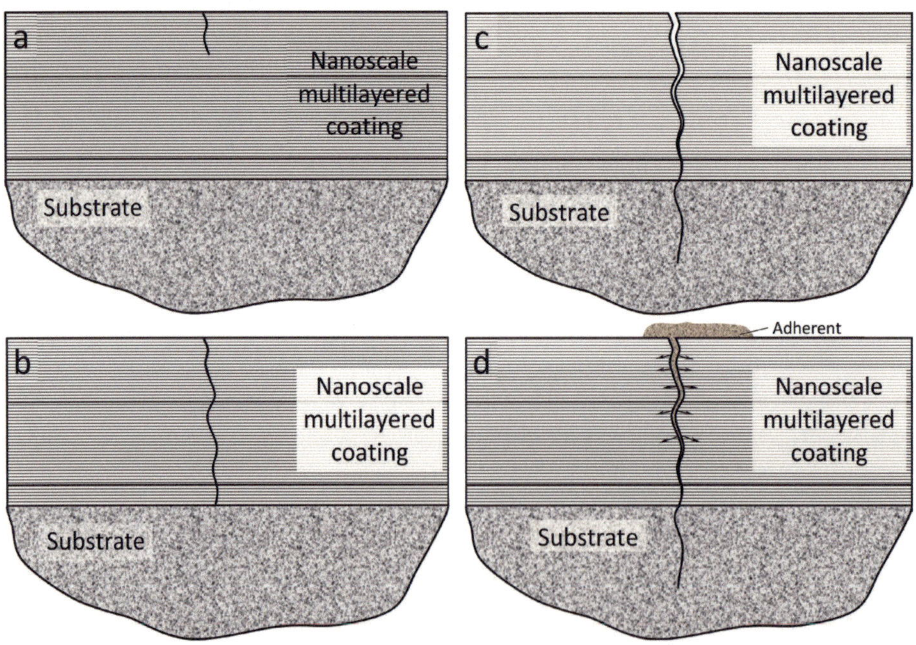

Figure 2. Stages of formation of a cross crack in multilayered nanostructured coatings: (a) initiation of a crack, (b) crack growth throughout the thickness of the coating, (c) "crack opening," (d) penetration into the crack of a fragment of the material being machined, resulting in "wedging" and increasing the width of the crack by up to 1 μm.

- Stage "a" is characterized by initiation of microcracks (see **Figure 2a**) caused by bending (usually alternating) mechanical and thermal stresses. An important role in initiation of a crack may be played by such concentrators of stresses as micro-drops (both embedded in the coating structure and superficial ones) (see **Figure 3a**), surface coating defects (craters, rough spots, etc.), inhomogeneity of mechanical properties of the coating, etc.

- Stage "b" is characterized by the growth of cracks, which can be inhibited by various obstacles, including: "bridging" of particles of more ductile phase, embedded in brittle phase; phase transformations of thermal nature or occurring as a result of plastic deformation in the development of crack tip; change in the direction of crack formation during passage of grain boundary [33].

- On the "c" stage, there is implementation of cracks in the carbide substrate, as well as its "disclosure"

- Stage "d" (typical only for a certain type of cracks) is characterized by penetration of particles into a crack of the material being machined (triggered by high temperatures and contact stresses resulting in thermoplastic deformation and yielding of the material), which has a "wedging" effect and stimulates further growth of the crack (see **Figure 4**). The studied crack development stage may cause separation of a fragment of coating or coated substrate (**Figure 5**). The examples of different cross cracks at different stages of development are shown in **Figures 3** and **4**.

The mechanism of wear of rake face of CI with NMCC Ti-TiN-(ZrNbTi)N has of complicated nature (see **Figure 5**). In particular, at the first stage of the wear process on rake face, the wear-fatigue processes with high frequency of oscillation of contact stresses prevail, and they considerably exceed the stresses acting on flank face of the tool and this results in weakening of local volumes of carbide material. The second stage marks the intensive weakening of local near-surface sections of rake face of CI in the area of crater formation and intensification of the processes of abrasive wear (**Figure 5**).

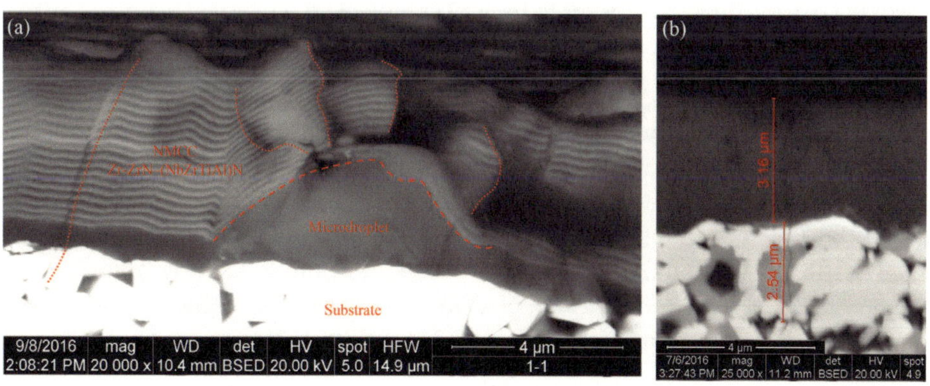

Figure 3. A microdroplet embedded into the structure of NMCC Cr-CrN-TiCrAlN as a factor, causing formation of a crack (a) and development of a cross crack in the coating TiN (b) [34].

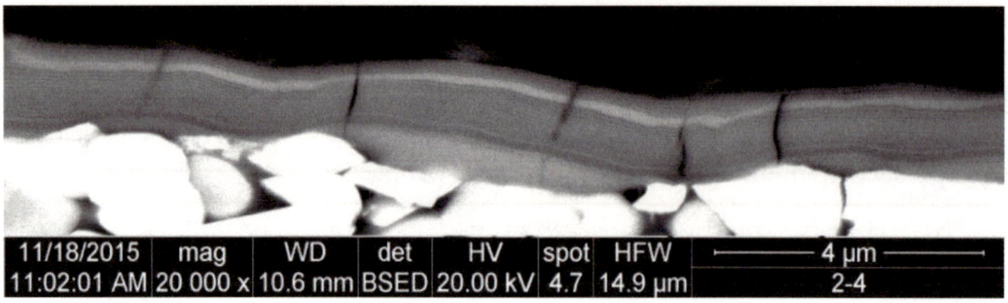

Figure 4. Example of formation of cross cracks in NMCC Cr-CrN-TiCrAlN (with crack thickness of 20–150 nm) [34].

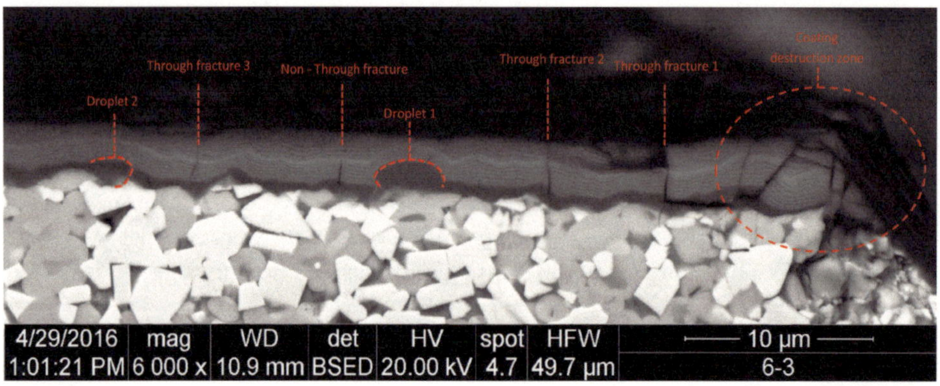

Figure 5. Specifics of wear and failure of NMCC Ti-TiN-(ZrNbTi) on rake face area between the cutting edge and the wear crater [33].

The data presented in **Figure 5** indicate the formation of through and dead-end cracks (1–3), as well as the presence of micro-droplets of cathode material, penetrating in the structure of the coating (1, 2), which contribute to the intensification of tool wear. The area of intense failure of the coating directly adjacent to the tool cutting edge can be characterized by mechanisms of abrasive and adhesion-fatigue wear, which result in active cracking.

Separation of a fragment of coated carbide substrate, occurring because of the final stage of the development of a cross crack is often accompanied by sufficient extensional chipping of the carbide substrate with locking in this area of a build-up of the material being machined (**Figure 6**). Such cracks can be formed in hard, but not sufficiently ductile NMCC, making it prone to brittle fracture. The maximum number of the formed cross-cracks was observed in NMCC Ti-TiAlN-TiAlN and Cr-CrN-TiCrAlN. The nature of such cracks was almost identical to the one usually observed for cracking of solid single-layered coatings of simple composition and architecture (for example, for coating TiN (see **Figure 3b**)).

In the formation of cross cracks in combination with longitudinal ones, three stages of crack development can be observed as follows (**Figure 7**). First, a longitudinal crack (separation) is formed between two fairly closely located longitudinal cracks that typically result in a separation of sufficiently large fragment of the coating (**Figure 8**). Next, the development of a cross

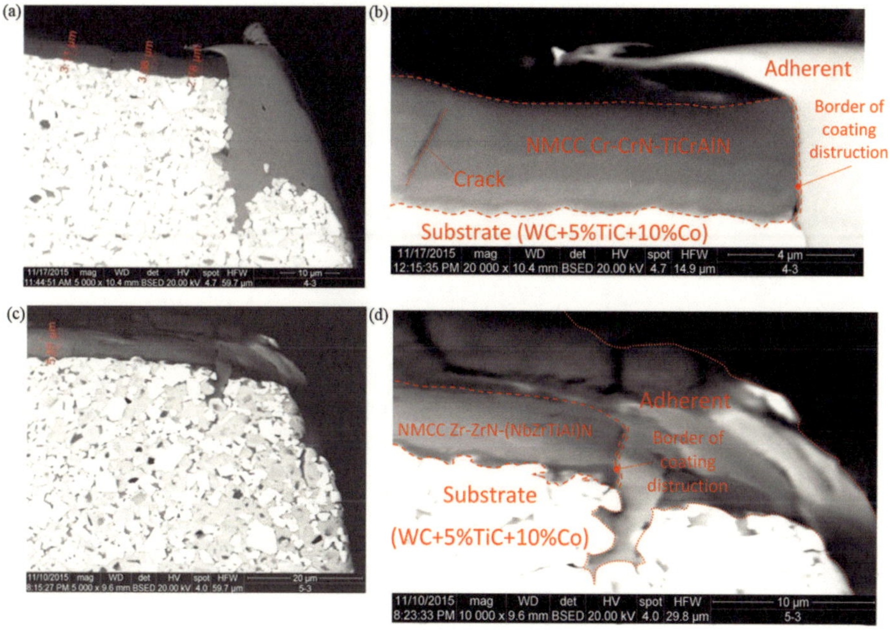

Figure 6. Examples of failure of coating and substrate because of breaking along the formed longitudinal crack. NMCC Cr-CrN-TiCrAlN (a, b); Zr-ZrN–(NbZrTiAl)N (c, d) [34].

crack is inhibited by its transformation into a longitudinal crack (delamination), which ceases its further development (**Figure 9b**, **c**). Finally, the growth of cross cracks that are periodically transformation into a longitudinal crack (delaminations) (**Figure 9a**), which is caused mainly by loss of their development energy.

It is noted that longitudinal cracks (delamination) (**Figure 10**) may occur in connection with violation of adhesive and cohesive bonds between layers and sublayers of the coating and in connection with the presence in the coating structure of different defects, resulting in the concentration of stresses (for example, micro-droplets (**Figure 11**), pores, failed "bridging" of adhesive bonds). The development of longitudinal cracks may result in the initiation of quite extensive internal delaminations without reaching the surface of the coating (this option is more favorable for tool life) and in output the cracks of the coating surface (**Figure 12**) with quite intensive subsequent damage of the coating and removal of the fragments of damaged NMCC by cut chips. The nanoscale layers of more ductile phase inhibit the development of cracks (**Figure 12b**) through formation of crack bridging.

To perform X-ray microanalysis using characteristic X-rays, emitted by the sample as a result of electron bombardment. The studies were conducted using a scanning electron microscope FEI Quanta 600 FEG. Following the analysis of the data presented in **Figures 13** and **14**, it is possible to note that the pattern of change in the content of the main chemical elements (N, O, Zr, Nb, Ti, W) in NMCC Ti-TiN-(ZrNbTi) relates to a significant growth of oxygen content and reduction of nitrogen content in the areas of through cracks in zones 1–3 (see **Figure 5**). Meanwhile, the increase in oxygen content and reduction in nitrogen content in the zones

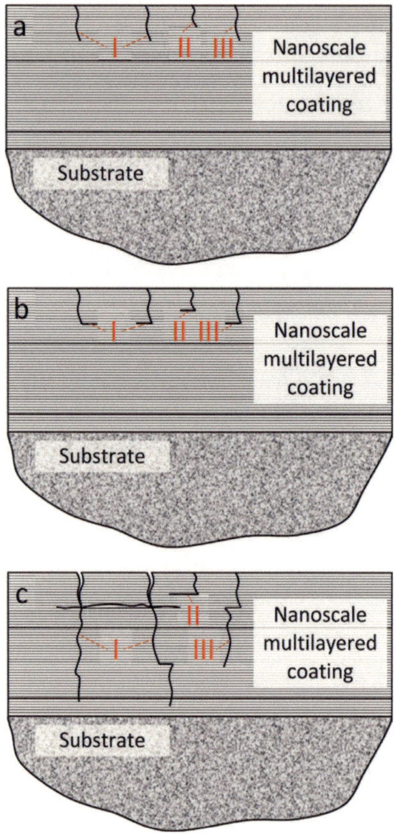

Figure 7. Stages of formation of a cross crack, transforming into a longitudinal crack (delamination) in multilayered nanostructured coating: (a) initiation of a crack, (b) transformation of a cross crack into a longitudinal crack (delamination); (c) creation of a through longitudinal crack between two cross cracks; (d) a tear-out or chipping of a coating segment between two cross cracks.

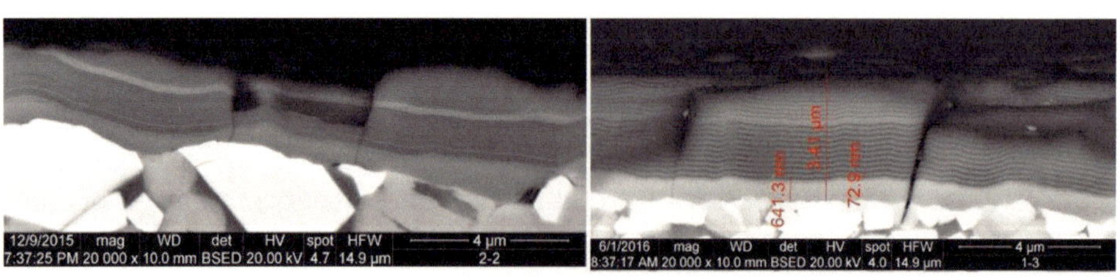

Figure 8. Chipping of coating fragments as a consequence of the formation of longitudinal cracks (delamination) between two cross cracks on the example of NMCC Zr-ZrN-(ZrCrAl)N [34].

through cracks is more noticeable for cracks located close to cutting edge of CI and coating failure zone. The increased oxygen content and reduced nitrogen content are also observed in the coating areas with a high content of microdroplets composed predominantly of α-Ti. Moreover, these processes are more visible for zones with the presence of larger microdroplets

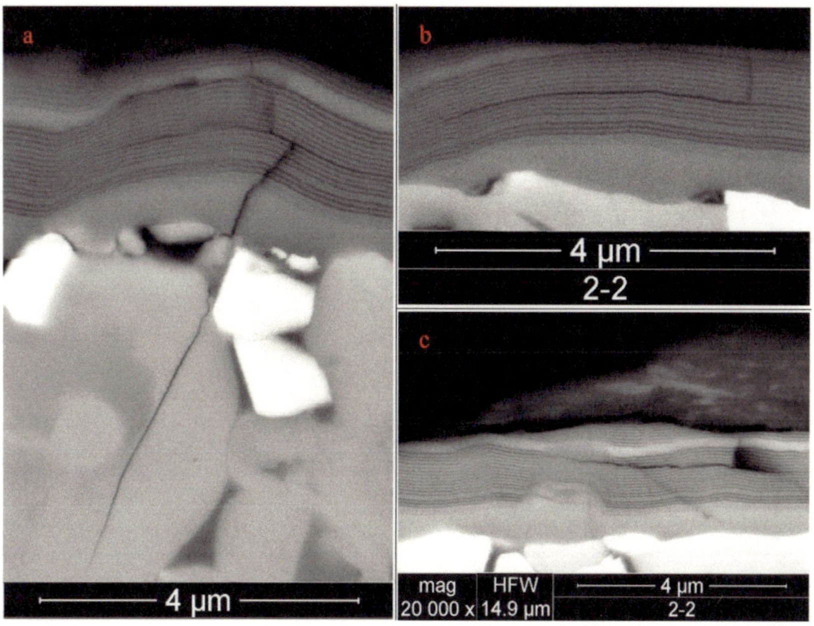

Figure 9. Transformation of cross cracks into longitudinal cracks (delamination) in NMCC Zr-ZrN-(ZrCrAl)N: partial transformation with continuous development of a cross crack (a) and complete transformation with stop of the growth of a cross crack (b, c) [34].

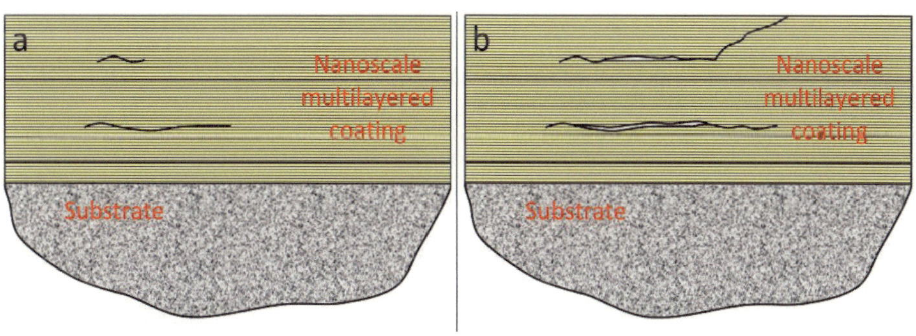

Figure 10. Stages of formation of a cross crack (delamination) in multilayered nanostructured coating: (a) initiation of a crack, (b) development of a crack.

(zone 1, see **Figure 5**). In zone I, associated with active cracking and failure of the coating, it is also possible to note growth of oxygen content and reduction of nitrogen content. Meanwhile, for the NMCC zones containing dead-end microcracks, slight change in the content of nitrogen and oxygen was observed.

The results of the analysis allow noting the partial dissociation of complex nitrides of NMCC and intensification of oxidative processes. Following the analysis of the data obtained, it is possible to note the following. The revealed sharp increase in oxygen content with significant reduction of nitrogen content indicates a high probability of intensive formation of solid and

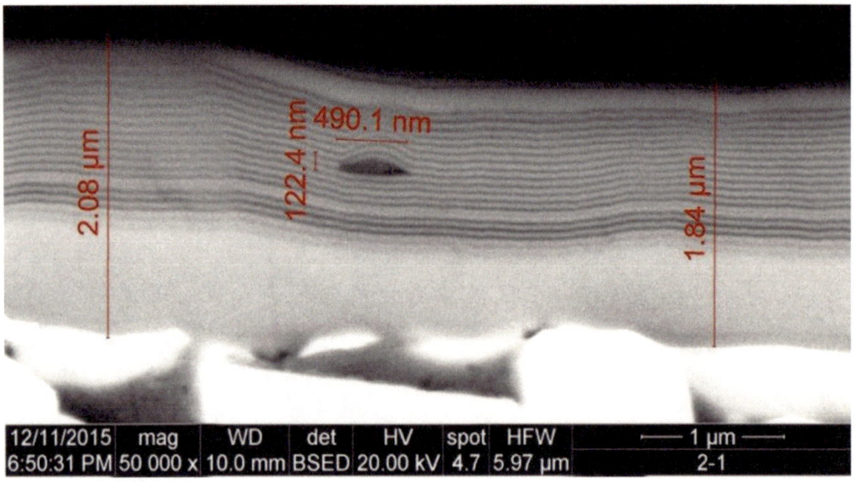

Figure 11. Internal defect of NMCC Zr-ZrN-(ZrCrAl)N (microdroplet), a possible cause of subsequent delamination.

relatively weak oxide formations of TiN_2, Al_2O_3 type, which failure can dramatically intensify the processes of abrasive wear of tool contact areas and reduce the efficiency of coatings deposited on carbide substrates. In this regard, in the use of the arc-PVD processes for the formation

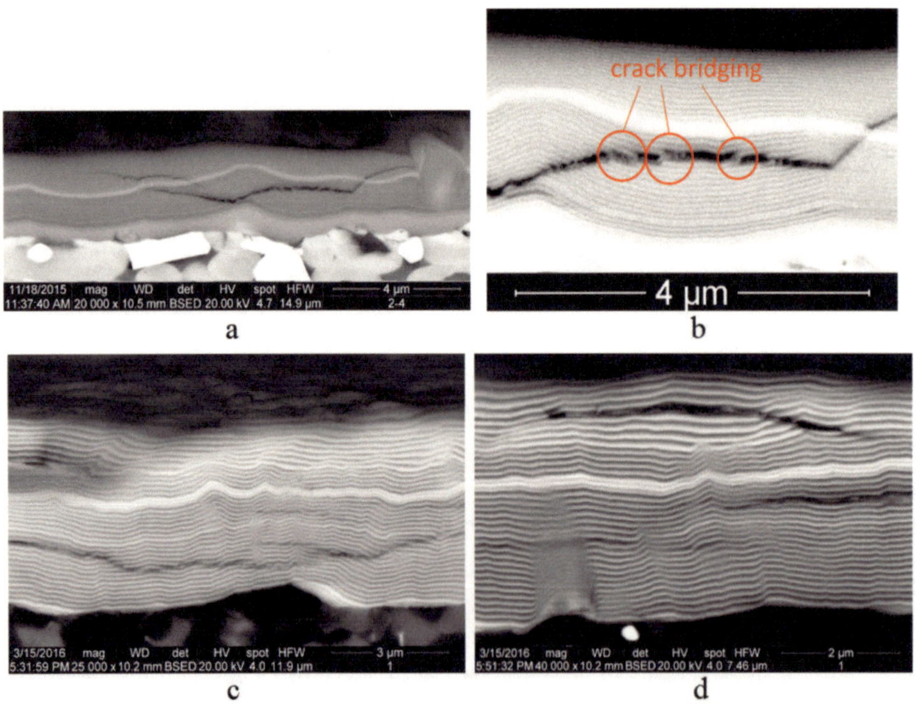

Figure 12. Examples of the formation of longitudinal cracks (delamination) NMCCZr-ZrN-(ZrCrAl)N (a, b) and Zr-ZrN-(ZrNbTiAl)N (c, d).

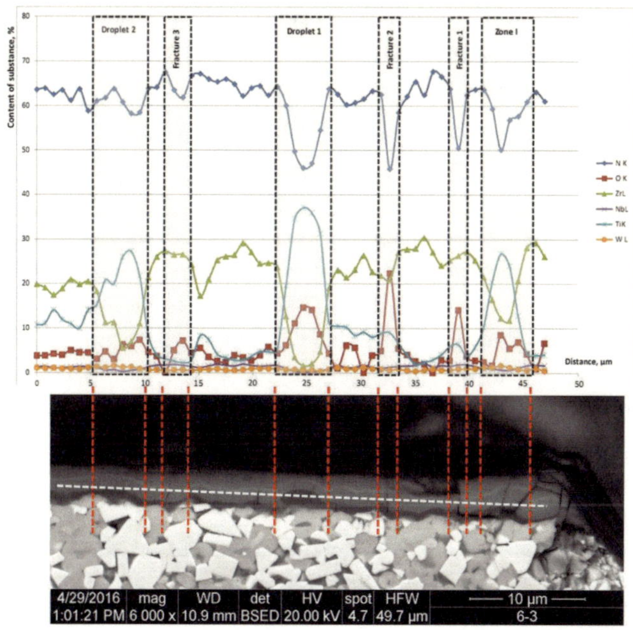

Figure 13. The results of the analysis of chemical element content, including of composition of NMCC Ti-TiN-(ZrNbTi) (N, O, Zr, Nb, Ti, W) at electron beam scanning length of 45 µm from the cutting edge of CI through the depth of NMCC, passing approximately equidistant from the carbide substrate and the outer boundary of NMCC [33].

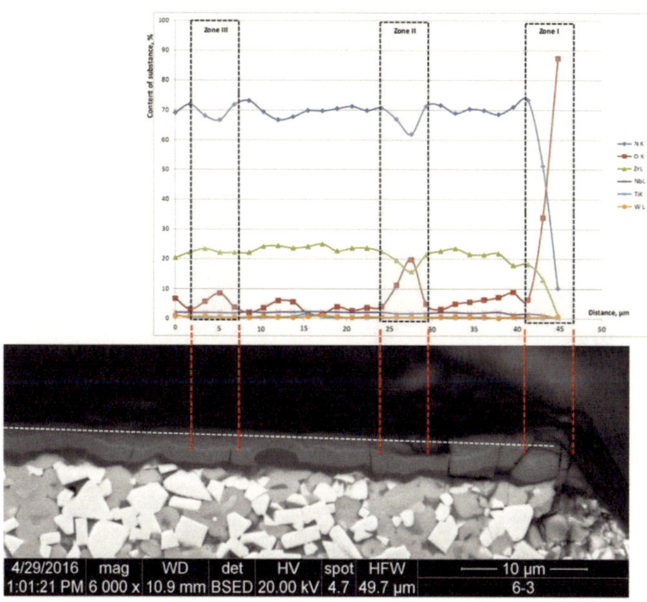

Figure 14. The pattern of change in the content of the main chemical elements (N, O, Zr, Nb, Ti, W) in NMCC Ti-TiN-(ZrNbTi) at the length of 45 µm occurring at the coating failure zone directly adjacent to the outer surface of the coating [33].

of wear-resistant coatings on carbide tools, it is necessary to use filtering systems that block the formation of macro- and microdroplets and that will result in significant improvement of the efficiency of coatings for various cutting operations.

4. Conclusions

Mechanisms of cracking in nanoscale multilayered composite coatings (NMCC) were considered. The studies were focused on NMCC of Zr-ZrN-(ZrCrAl)N, Cr-CrN-(TiCrAl)N, and Zr-ZrN-(ZrNbTiAl)N and Ti-TiN-(ZrNbTi) deposited through FCVAD technology. The thickness of the NMCC under study reached 2.44–11.7 μm. An uncoated carbide tool and a carbide tool with a "reference" TiN coating were selected as objects for comparison. Preliminary studies have shown that the NMCCs under study are characterized by high adhesion to substrate and high hardness (34–38 GPa). The tests of cutting properties carried out at longitudinal turning of steel C45 (HB 200) at the following cutting modes: $f = 0.2$ mm/rev; $a_p = 1.0$ mm; $v_c = 250$ m min^{-1} have shown that all NMCC under study sufficiently improve tool life (by 3–4 times in comparison with uncoated tool, and by 1.5–2 times in comparison with the tool with the TiN "reference" coating). Meanwhile, the nature of wear of NMCC with harder and less ductile wear-resistant layers ((TiCrAl)N and (TiAl)N) was considerably different from the mechanism of wear of NMCC with less hard and more ductile wear-resistant layers, which included zirconium nitrides ((ZrCrAl) N and (ZrNbTiAl)N) compositions. The tools with harder and less ductile NMCC showed less steady kinetics of wear, characterized by sharp intensification of wear and failure in transition from "steady" to catastrophic wear, i.e., at the end of the tool life period. The microstructural studies carried out at cross-cut sections of carbide samples with the NMCC under study with the use of a SEM revealed the nuances and mechanisms of formation of different types of cracks in the various NMCCs studied.

The X-ray microanalysis showed the considerable increase in oxygen content and reduction in nitrogen content in the following coating zones: (i) areas of through transverse cracks in the coating; and (ii) areas adjacent to micro-droplets embedded in coating structure.

Acknowledgments

This study was supported by Grant of the Russian Science Foundation (theme № 15-36/RNF Agreement № 15-19-00231 from "18" May 2015).

Nomenclature

FCVAD - filtered cathodic vacuum arc deposition

CF – cutting fluid

http://dx.doi.org/10.5772/68008

PVD - physical vapour deposition

NMCC - nanostructured multi-layered composite coating

Author details

Alexey A. Vereschaka[1]*, Sergey N. Grigoriev[1], Nikolay N. Sitnikov[2] and Gaik V. Oganyan[1]

*Address all correspondence to: ecotech@rambler.ru

1 Moscow State Technological University "Stankin", Moscow, Russia

2 Federal State Unitary Enterprise "Keldysh Research Center", Moscow, Russia

References

[1] E. Harry, M. Ignat, Y. Pauleau, A. Rouzaud, P. Juliet. Mechanical behavior of hard PVD multilayered coatings. Surface and Coatings Technology. 2000;**125**:185-189.

[2] J.L. Mo, M.H. Zhu, A. Leyland, A. Matthews. Impact wear and abrasion resistance of CrN, AlCrN and AlTiN PVD coatings. Surface & Coatings Technology. 2013;**215**: 170-177.

[3] Y. Birol. Sliding wear of CrN, AlCrN and AlTiN coated AISI H13 hot work tool steels in aluminium extrusion. Tribology International. 2013;**57**:101-106.

[4] J. Nohava, P. Dessarzin, P. Karvankova, M. Morstein. Characterization of tribological behavior and wear mechanisms of novel oxynitride PVD coatings designed for applications at high temperatures. Tribology International. 2015;**81**:231-239.

[5] L. Aihua, D. Jianxin, C. Haibing, C. Yangyang, Z. Jun. Friction and wear properties of TiN, TiAlN, AlTiN and CrAlN PVD nitride coatings. International Journal of Refractory Metals and Hard Materials. 2012;**31**:82-88.

[6] M. Antonov, I. Hussainova, F. Sergejev, P. Kulu, A. Gregor. Assessment of gradient and nanogradient PVD coatings behaviour under erosive, abrasive and impact wear conditions. Wear. 2009;**267**:898-906.

[7] P. Henry, M.J. Pac, C. Rousselot, M.H. Tuilier. Wear mechanisms of titanium and aluminium nitride coatings: a microtribological approach. Surface & Coatings Technology. 2013;**223**:79-86.

[8] B.D. Beake, J.F. Smith, A. Gray, G.S. Fox-Rabinovich, S.C. Veldhuis, J.L. Endrino. Investigating the correlation between nano-impact fracture resistance and hardness/modulus ratio from nanoindentation at 25-500°C and the fracture resistance and lifetime of cutting tools with Ti1−xAlxN (x = 0.5 and 0.67) PVD coatings in milling operations. Surface & Coatings Technology. 2007;**201**:4585-4593.

[9] G. Skordaris, K.D. Bouzakis, P. Charalampous. A dynamic FEM simulation of the nano-impact test on mono- or multi-layered PVD coatings considering their graded strength properties determined by experimental–analytical procedures. Surface & Coatings Technology. 2015;**265**:53-61.

[10] B.D. Beake, Li Ning, Ch. Gey, S.C. Veldhuis, A. Komarov, A. Weaver, M. Khanna, G.S. Fox-Rabinovich. Wear performance of different PVD coatings during hard wet end milling of H13 tool steel. Surface & Coatings Technology. 2015; 279: 118-125 doi: 10.1016/j.surfcoat.2015.08.038

[11] L. Ning, S.C. Veldhuis, K. Yamamoto. Investigation of wear behavior and chip formation for cutting tools with nano-multilayered TiAlCrN/NbN PVD coating. International Journal of Machine Tools & Manufacture. 2008;**48**:656-665.

[12] G.S. Fox-Rabinovich, K. Yamamoto, A.I. Kovalev, S.C. Veldhuis, L. Ning, L.S. Shuster, A. Elfizy. Wear behavior of adaptive nano-multilayered TiAlCrN/NbN coatings under dry high performance machining conditions. Surface & Coatings Technology. 2008;**202**: 2015-2022.

[13] G.S. Fox-Rabinovich, A.I. Kovalev, M.H. Aguirre, B.D. Beake, K. Yamamoto, S.C. Veldhuis, J.L. Endrino, D.L. Wainstein, A.Y. Rashkovskiy. Design and performance of AlTiN and TiAlCrN PVD coatings for machining of hard to cut materials. Surface & Coatings Technology. 2009;**204**:489-496.

[14] G. Erkens, R. Cremer, T. Hamoudi, K.D. Bouzakis, J. Mirisidis, S. Hadjiyiannis, G. Skordaris, A. Asimakopoulos, S. Kombogiannis, J. Anastopoulos. Super nitrides: a novel generation of PVD hard coatings to meet the requirements of high demanding cutting applications. CIRP Annals–Manufacturing Technology. 2003;**52**(1):65-68.

[15] B.D. Beake, G.S. Fox-Rabinovich. Progress in high temperature nanomechanical testing of coatings for optimizing their performance in high speed machining. Surface & Coatings Technology. 2014;**255**:102-111.

[16] H. Ezuraa, K. Ichijoa, H. Hasegawab, K. Yamamotoc, A. Hottaa, T. Suzukia. Microhardness, microstructures and thermal stability of (Ti, Cr, Al, Si)N films deposited by cathodic arc method. Vacuum. 2008;**82**:476-481.

[17] C.H. Lai, K.H. Cheng, S.J. Lin, J.W. Yeh. Mechanical and tribological properties of multi-element (AlCrTaTiZr)N coatings. Surface & Coatings Technology. 2008;**202**:3732-3738.

[18] M.G. Faga, G. Gautier, R. Calzavarini, M. Perucca, E. Aimo Boot, F. Cartasegna, L. Settineri. AlSiTiN nanocomposite coatings developed via Arc Cathodic PVD: evaluation of wear resistance via tribological analysis and high speed machining operations. Wear. 2007;**263**:1306-1314.

[19] A.A. Vereschaka, M.A. Volosova, S.N. Grigoriev, A.S. Vereschaka. Development of wear-resistant complex for high-speed steel tool when using process of combined cathodic vacuum arc deposition. Procedia CIRP. 2013;**9**:8-12.

http://dx.doi.org/10.5772/68008

[20] A.A. Vereschaka, M.A. Volosova, A.D. Batako, A.S. Vereshchaka, B.Y. Mokritskii. Development of wear-resistant coatings compounds for high-speed steel tool using a combined cathodic vacuum arc deposition. International Journal of Advanced Manufacturing Technology. 2016;**84**:1471-1482.

[21] S.N. Grigoriev, A.A. Vereschaka, A.S. Vereschaka, A.A. Kutin. Cutting tools made of layered composite ceramics with nano-scale multilayered coatings. Procedia CIRP. 2012; **1**:318-323.

[22] A.A. Vereshchaka, A.S. Vereshchaka, O. Mgaloblishvili, M.N. Morgan, A.D. Batako. Nano-scale multilayered-composite coatings for the cutting tools. International Journal of Advanced Manufacturing Technology. 2014;**72**(1):303-317.

[23] K.D. Bouzakis, N. Michailidis, G. Skordaris, E. Bouzakis, D. Biermann, R. M'Saoubi. Cutting with coated tools: coating technologies, characterization methods and performance optimization. CIRP Annals—Manufacturing Technology. 2012;**61**:703-723.

[24] V.P. Tabakov. The influence of machining condition forming multilayer coatings for cutting tools. Key Engineering Materials. 2012;**496**:80-85.

[25] V.P. Tabakov, A.V. Chikhranov. Research of developments of cracks in wearproof coverings of cutting tool. Vestnik TSU. 2013;**18**(4):1916-1917 (in Russian).

[26] V.P. Tabakov, M.Y. Smirnov, A.V. Tsirkin. Productivity of end mills with multilayer wear-resistant coatings. Ulyanovsk: UlSTU. 2005 (in Russian).

[27] A.A Vereschaka, A.S. Vereschaka, J.I. Bublikov, A.Y. Aksenenko, N.N. Sitnikov. Study of properties of nanostructured multilayer composite coatings of Ti-TiN-(TiCrAl)N and Zr-ZrN-(ZrNbCrAl)N. Journal of Nano Research. 2016;**40**:90-98.

[28] A.A. Vereschaka. Development of assisted filtered cathodic vacuum arc deposition of nano-dispersed multi-layered composite coatings on cutting tools. Key Engineering Materials. 2014;**581**:62-67.

[29] A.O. Volkhonskii, A.A. Vereshchaka, I.V. Blinkov, A.S. Vereshchaka, A.D. Batako. Filtered cathodic vacuum Arc deposition of nano-layered composite coatings for machining hard-to-cut materials. International Journal of Advanced Manufacturing Technology. 2016;**84**:1647-1660.

[30] S. Kumar, W.A. Curtin. Crack interaction with microstructure. Materials Today. 2007;**10**(9):34-44.

[31] W. Hume-Rothery, Atomic theory for students of metallurgy. London: The Institute of Metals. 1969 (fifth reprint).

[32] W.C. Oliver, G.M.J. Pharr. An improved technique for determining hardness and elastic modulus using load and displacement sensing indentation. Journal of Materials Research. 1992;**7**:1564-1583.

[33] A.A. Vereschaka, B.Y. Mokritskii, N.N. Sitnikov, G.V. Oganyan, A.Y. Aksenenko. Study of mechanism of failure and wear of multi-layered composite nano-structured coating based on system Ti-TiN-(ZrNbTi)N deposited on carbide substrates. Journal of Nano Research. 2017;**45**:110-123.

[34] A.A. Vereschaka, S.N. Grigoriev. Study of cracking mechanisms in multi-layered composite nano-structured coatings. Wear. 2017;**378-379**:43-57. DOI:10.1016/j.wear.2017.01.101.

Section 2

Fabrication, Characterization and Applications

Chapter 5

Gd-Doped Superparamagnetic Magnetite Nanoparticles for Potential Cancer Theranostics

Maheshika Palihawadana-Arachchige,
Vaman M. Naik, Prem P. Vaishnava,
Bhanu P. Jena and Ratna Naik

Additional information is available at the end of the chapter

http://dx.doi.org/10.5772/intechopen.68219

Abstract

Nanotechnology has facilitated the applications of a class of nanomaterials called superparamagnetic iron oxide nanoparticles (SPIONs) in cancer theranostics. This is a new discipline in biomedicine that combines therapy and diagnosis in one platform. The multifunctional SPIONs, which are capable of detecting, visualizing, and destroying the neoplastic cells with fewer side effects than the conventional therapies, are reviewed in this chapter for theranostic applications. The chapter summarizes the design parameters such as size, shape, coating, and target ligand functionalization of SPIONs, which enhance their ability to diagnose and treat cancer. The review discusses the methods of synthesizing SPIONs, their structural, morphological, and magnetic properties that are important for theranostics. The applications of SPIONs for drug delivery, magnetic resonance imaging, and magnetic hyperthermia therapy (MHT) are included. The results of our recent MHT study on Gd-doped SPION as a possible theranostic agent are highlighted. We have also discussed the challenges and outlook on the future research for theranostics in clinical settings.

Keywords: theranostics, Fe_3O_4 nanoparticles, MRI contrast agent, drug delivery, magnetic hyperthermia

1. Introduction

Nanomaterials, with the size of at least one dimension ranging from a few nanometers to about a hundred nanometers, having unique properties compared to their respective bulk materials, are of intense research interest because of their applications in various fields of science and technology. One of the major applications, among many of their potential applications, is in biomedicine as platform for effective diagnosis and therapy [1–3]. The multifunctionality of these nanoparticles has recently led the biomedicine research in a new direction called "Theranostics" which is the integration of diagnostic imaging and therapeutic function into a single platform [4, 5]. Theranostic agents allow the combination of diagnosis, treatment, and follow-up of a disease and hence are expected to contribute to personalized medicine. Among many nanomaterials, magnetic nanoparticles (MNPs) have the potential to deliver imaging and therapeutic agents to a specific region in the body with an external magnetic field manipulation. This requires large magnetization for the MNP so that they could respond to externally applied magnetic fields at physiological temperatures. Superparamagnetic iron oxide nanoparticles (SPIONs), such as Fe_3O_4 and γ-Fe_2O_3 nanoparticles, exhibit relatively higher saturation magnetization with no magnetic hysteresis (zero remanence and coercivity) and fulfill other major requirements such as low toxicity, biocompatibility, and surface functionalization capabilities for theranostic applications. A number of SPIONs have undergone clinical trials and several formulations have been approved for clinical imaging and therapeutic applications [6]. A few examples are Lumiren for bowel imaging, Ferridex IV for liver and spleen imaging, Combidex for lymph node metastases imaging, and Ferumoxytol for iron deficiency therapy.

Furthermore, SPIONs can be multipurposely used for diagnosis such as magnetic resonance imaging (MRI) and for therapeutic functions such as targeted delivery of therapeutic agents, anticancer drugs, siRNA, and for magnetic hyperthermia (MHT) for cancer treatments. This makes SPION an ideal vehicle in the development of theranostic nanomedicine [7–9]. An example of strategy for using magnetic nanoparticles as a potential theranostic agent is illustrated in **Figure 1**.

In this chapter, we discuss the detailed background on magnetic properties of SPIONs and their synthesis methods and surface modification for cancer diagnosis and therapy. In addition, various applications of SPION ranging from MRI contrast agent to therapeutic-targeted drug delivery and MHT are discussed. We have also highlighted the results of our recent study on Gd-doped SPION as a possible theranostic agent. The remainder of the chapter focuses on the challenges and outlook on the future research for theranostics in clinical settings.

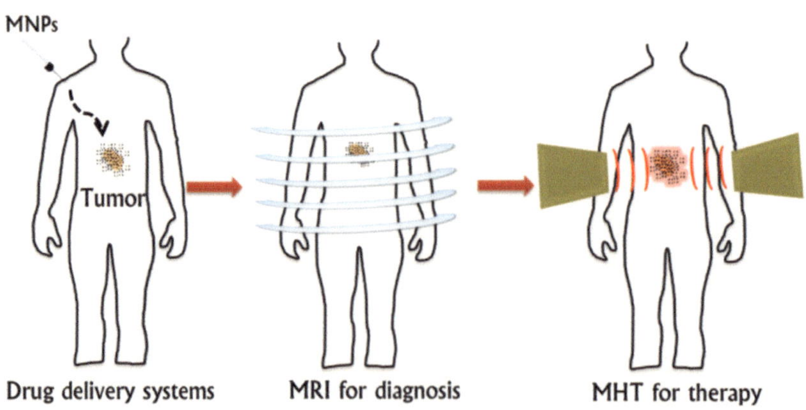

Figure 1. Schematic illustration of the therapeutic strategy using MNP. Functionalized MNPs accumulate in the tumor tissues via the drug delivery system. MNP can be used as a tool for cancer diagnosis by MRI or for magneto-impedance sensor. Hyperthermia can then be induced by alternating magnetic field exposure.

http://dx.doi.org/10.5772/intechopen.68219

2. Magnetic properties of SPION

2.1. Background

MNPs have been studied for over 50 years now due to their potential application in many areas including biomedical sciences. As for the types of MNP, the major focus has been on iron oxide (Fe_3O_4), gold-coated iron oxide ($Au\text{-}Fe_3O_4$), metallic iron (Fe), and Fe-Co and Fe-Pt nanoparticles. In most cases, the particle size ranges from 1 to 100 nm exhibiting high surface-to-volume ratio. As a result, they offer higher surface area for interaction with foreign objects compared to larger particles. Many review articles have been written that focus on sensing, drug delivery, and hyperthermia properties of these nanoparticles [10–13]. The physical, chemical, and magnetic properties of MNP largely depend on synthesis method and their surface modification, and much progress has been made in this direction to MNP of varying sizes, shapes, composition, and core-shell designs [14–23].

The important magnetic parameters relevant to theranostic applications are saturation magnetization (M_s), remanent magnetization (M_r), coercivity (H_c), Curie temperature (T_c), magnetic anisotropy energy density (K), and blocking temperature (T_b). These parameters are influenced by the material, size, shape, composition, and core-shell (functionalization) of the nanoparticles. M_s is the maximum value of magnetization of the material that can be achieved under the influence of an external magnetic field, M_r is the remanent magnetization in the material after removing the external magnetic field, H_c is the strength of the reverse magnetic field needed to bring the remanent magnetization to zero, and K is the material property signifying the tendency of the magnetization to orient along a certain axis of the particle. As the volume (V) of the particles decreases, the magnetic anisotropy energy (KV) of the nanoparticle also decreases. If the particle size is reduced below a certain critical size, it becomes a single magnetic domain creating a giant spin called "superspin" leading to a large magnetic moment (~10,000 Bohr magneton) on each particle. The behavior of a collection of such noninteracting particles under an external magnetic field is determined by a competition between the magnetic anisotropy energy barrier (ΔE) and the thermal energy ($k_B T$) for magnetic moment reversal. Above a characteristic temperature called the blocking temperature, T_D, their behavior is very similar to that of a paramagnetic material and described as "superparamagnetism." The underlying physics of superparamagnetism is founded on the activation law for the relaxation time τ of the net magnetization of the particle given by $\tau = \tau_o \exp(\Delta E/k_B T)$, where τ_o is of the order of 10^{-9}–10^{-12} s [24].

2.2. Effect of size, shape, and composition

Magnetic properties of materials, such as susceptibility, coercivity, and saturation magnetization, depend on the structure, size, shape, and composition, and can be altered to manipulate the magnetic properties. Particle size plays an important role in many magnetic biomaterial applications such as magnetic hyperthermia and drug delivery, where the size used lies in the nanometer regime. The MNPs often contain a layer of disordered spins on the surface of the particle leading to reduction in their M_s compared to the corresponding bulk material. A relation between the M_s and the size of the nanoparticle is given by [25]

$$M_s = M_{sb}\left(\frac{r-d}{r}\right)^3 \tag{1}$$

where r is the radius of the nanoparticle, d is the layer thickness of the disordered spins, and M_{sb} is the saturation of the bulk material. Recent studies have shown that the functionalization of MNP can reduce the thickness of the surface-disordered spin layer [26].

Although the effect of shape of MNP on their magnetic properties is not extensively studied, a few investigations have been reported in the literature on ferrite nanocubes, maghemite nanorods, NiFe wires, cobalt nanodiscs, tetrapods, and Au-MnO nanoflowers showing a strong dependence of M_s on the shapes of the nanoparticles [27–36]. Higher M_s values have been observed for the cubic MNP compared to the spherical MNP of the same size [37]. Also, cubic Fe_3O_4 nanoparticles have been found to exhibit higher T_B compared to spherical Fe_3O_4 nanoparticles [38], and the amount of disordered spins to be less (4%) in the former and more (8%) in the latter [39].

Magnetic properties of widely used magnetite (Fe_3O_4) with its spinel structure of $[Fe^{3+}]_A[Fe^{3+}Fe^{2+}]_BO_4$ can be changed if other magnetic atoms such as Ni, Co, Mn, and so on are substituted at the tetrahedral A or octahedral B sites of the spinel structure. This flexibility of creating mixed ferrites is useful in tuning the magnetic properties for hyperthermia applications. There have been numerous studies investigating the interdependence of magnetic properties and the composition. The method of preparation, concentration and nature of dopants, and postsynthesis processes have shown to profoundly affect the magnetic properties. A study [40] compared the magnetization among the four spinels of $FeFe_2O_4$, $MnFe_2O_4$, $CoFe_2O_4$, and $NiFe_2O_4$ for the same size of 12 nm and found the highest magnetization for $MnFe_2O_4$. In another study with $Y_3Fe_{5-x}Al_xO_{12}$ for x varying between 0 and 2, the Curie temperature changed from 40 to 280°C [41]. With increasing Al, Fe^{3+} cations occupied the tetrahedral sites and some of the octahedral sites of Fe^{2+} were replaced by nonmagnetic Al^{3+} cations, which reduced the magnitude of M_s. The T_c value reached the room temperature for x value between 1.5 and 1.8. The variation in composition affects not only the magnitude of M_s but also the coercivity. The tailoring of ferromagnetic to paramagnetic phase transition temperature is particularly very useful in hyperthermia application to turn off undesirable heating beyond the required temperature.

3. SPION synthesis and surface modification

Over the past decades, many efficient synthesis methods have been developed to produce the size/shape controlled, stable, biocompatible, and monodispersed iron oxide nanoparticles [42–45]. The most common methods include coprecipitation [46, 47], thermal decomposition [48], hydrothermal synthesis [49, 50], microemulsion [51], and sonochemical [52] synthesis. Thermal decomposition technique involves decomposition of organo-metallic iron precursors in organic solvents at higher temperatures. Although the method can produce high-quality monodisperse particles because of separate nucleation and growth processes, it is a complicated synthesis method and produces hydrophobic nanoparticles that cannot be directly used for bio-applications without laborious postsynthesis processes, which may result in aggregation and loss of magnetic properties. The most commonly used technique is the coprecipitation method, which is a cost-effective and a facile synthesis method. However, this method produces Fe_3O_4 nanoparticles with wide particle size distribution due to lack of control over hydrolysis reactions of the iron precursors, and the nucleation and growth steps leading to particles with a range of superparamagnetic-blocking temperature. The other common, recently developed, method is the hydrothermal synthesis, which generates nanoparticles with excellent crystallinity with controllable size and shape in aqueous phase. The properties of the nanoparticles can vary with the synthesis method due to the differences in cationic distribution and vacancies, spin canting, or surface contribution.

In the design of magnetic nanoparticles for theranostic applications, surface modification plays an important role in providing colloidal stability and biocompatibility. The stable colloidal suspensions of surfactant-coated SPION are called "ferrofluids" which are magnetizable and remain as liquids in the presence of magnetic fields and in biological media. Stabilization of the ferrofluid occurs in the presence of one or both of the two repulsive forces (see **Figure 2**). The electrostatic repulsion can be understood through the knowledge of the diffusion potential and mainly depends on the ionic strength and the pH of the solution. The steric force is difficult to predict or quantify and mostly depends on the weight and the density of the polymer used for the coating.

In order to achieve biocompatibility, the coating should prevent any toxic ion leakage from magnetic core into the biological environment as well as shielding the magnetic core from oxidation and corrosion. When nanoparticles are injected into the body during in vivo applications, they are often recognized by reticuloendothelial system (RES) that eliminates any foreign substance from blood stream [53]. In this process, nanoparticles are rapidly attacked by the plasma proteins from RES and shuttled out of circulation to the liver, spleen, or kidney,

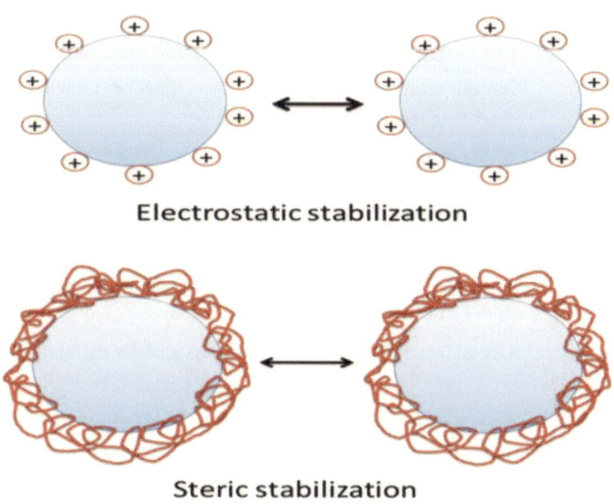

Figure 2. Electrostatic and steric repulsion between the particles.

which are then cleared out from the body. Also, this RES accumulation often causes toxicity issues as well. The specific surface coatings can prevent the adsorption of these proteins, increasing the circulation time in blood, hence maximizing the possibility to reach target tissues [54]. For instance, it is well known that coating of hydrophilic polymers, mainly polyethylene glycol (PEG), on the nanoparticles reduces nonspecific binding of the proteins resulting in stealth behavior. In addition to the stabilization and enhanced biocompatibility, these protecting shells also provide a platform for further functionalization such as the addition of specific targeting ligands, dyes, or therapeutic agents.

Over the years, researchers have developed various surface modification strategies composed of grafting of or coating with both organic and inorganic materials. Organic molecules include small organic molecules, macromolecules or polymer and biological molecules. They provide various highly reactive functional groups such as carboxyl groups, aldehyde groups, and amino groups. Polymer-coating materials can be classified into synthetic and natural, and some commonly used polymers are listed in **Table 1** along with their advantages.

The surface coating could affect the magnetic properties of SPION. Many studies have reported the effect of the surfactants on the magnetic properties [55–60]. Yuan et al. [58] investigated the effect of surfactant on magnetic properties using commercially available aqueous nanoparticle suspensions, FluidMAG-Amine, FluidMAG-UC/A, and FluidMAG-CMX, in parallel with oleic acid-covered particles suspended in hexane and heptane. Their results reveal the reduction of magnetic phase in nanoparticles, which varies with different coatings as well as with solvents. The reduction in magnetization with different coatings was attributed to the different degree of surface spin disorder.

Polymer		Advantages and applications	References
Natural	Dextran	Stability, biocompatibility, enables optimum polar interactions with iron oxide surfaces, and enhances the blood circulation time	[61–66]
	Starch	Improves the biocompatibility, good for MRI, and drug target delivery	[67, 68]
	Chitosan	Biocompatible and hydrophilic large abundance in nature, biocompatibility, and ease of functionalization. widely used as nonviral gene delivery system	[69–72]
Synthetic	Poly(ethylene-glycol) (PEG)	Enhances the hydrophilicity and water-solubility, improves the biocompatibility, blood circulation times, and internalization efficiency of the nanoparticles. Used in target-specific cell labeling, magnetic hyperthermia, targeted drug delivery	[73–75]
	Alginate	Improves the stability and biocompatibility. Used in drug delivery applications	[76–78]
	Poly-N-isopropyl-acrylamide (PNIPAM)	Generally used as thermosensitive drug delivery and cell separation	[79–82]
	Polyethylene-imine (PEI)	Ability to complex with DNA, guide intracellular trafficking of their cargo into the nucleus, used for gene delivery cell transfection with either DNA or siRNA nucleotides	[83–86]

Table 1. Commonly studied organic polymers and their advantages.

4. SPION for drug delivery

Over the last two decades, MNPs have been increasingly exploited as platforms for the transport of therapeutics including drugs and genes [46, 87, 88]. In magnetic drug delivery, a drug or a therapeutic reagent is conjugated to the nanoparticle and introduced in the body, and concentrated in the target area by means of a magnetic field gradient (using an internally implanted permanent magnet or an externally applied field) [89]. Even though magnetic drug delivery shows a great promise in cancer treatment avoiding the side effects of conventional chemotherapy, the designing and fabrication of an efficient nanoparticle-based drug delivery system is still a challenge. Using a targeting ligand, the targeting specificity can be enhanced. These anticancer drugs carried by the nanoparticles can then be released at the tumor site via enzymatic activity, or via changes in the physiological conditions such as temperature and pH. Drug release can also be magnetically triggered from the drug-conjugated magnetic nanoparticles [89–91]. For example, Hayashi et al. [92] reports a study done on superparamagnetic iron oxide nanoparticles conjugated with folic acid (well known as a targeting ligand for breast cancer cells), β-Cyclodextrin (which acts as drug container), and Tamoxifen (anticancer drug). Using an AC magnetic field, heat is generated which triggers drug release—a behavior that is controlled by switching the high-frequency magnetic field on and off. This is capable of performing drug delivery and hyperthermia simultaneously. Among various other anticancer drugs, Doxorubicin (Dox) is widely used as a model drug. There are several methods that can be used to load Dox into nanoparticles, such as by adsorption onto nanocarrier inorganic core [93–95], by diffusion [95, 96], or entrapment [97, 98] in the coating materials and by chemical bonds [99, 100] with the coating of the nanocarrier. Several modifications including surface functionalization of these SPIONs with Dox have been conducted over the last few years to investigate their efficacy [101]. Previous studies have reported that PEG-functionalized porous silica shell onto Dox-conjugated Fe_3O_4 nanoparticle cores [102], PAMAM (Poly(amidoamine))-coated Fe_3O_4 nanoparticles-Dox complex [103], and Dox-loaded Fe_3O_4 nanoparticles modified with PLGA-PEG copolymers [104] could potentially be very promising in therapeutic cancer treatment. However, most of the Dox-SPION–based drug delivery studies have been focused on human breast cancer cells.

In our recently published work [105], we have developed a novel drug delivery platform based on Fe_3O_4 nanoparticles as a vehicle for an anticancer drug (Dox), attached to a model dye (FITC) for their precise tracking and investigated their incorporation into the human pancreatic cancer cell line (MIA PaCa-2) for specific drug targeting. Existing EDC/NHS technique was employed for this dual drug/dye conjugation. This unique drug-dye dual conjugation of SPION after penetration through the cell membrane shows a steady release of Dox into the nucleus of the malignant cells. Our studies demonstrate that the association of Dox onto the surface of nanoparticles enhances its penetration into the cancer cells as compared to the unconjugated drug as shown in the subsequent text (**Figure 3**). In addition to the rapid uptake of these SPIONs by live cells, our results also suggest that upon entering the cells, Dox is cleaved from the conjugation, which might be due to the enzymatic reactions that occur within the cells, and tends to accumulate in the nuclei fulfilling the major requirement for an effective therapeutic system.

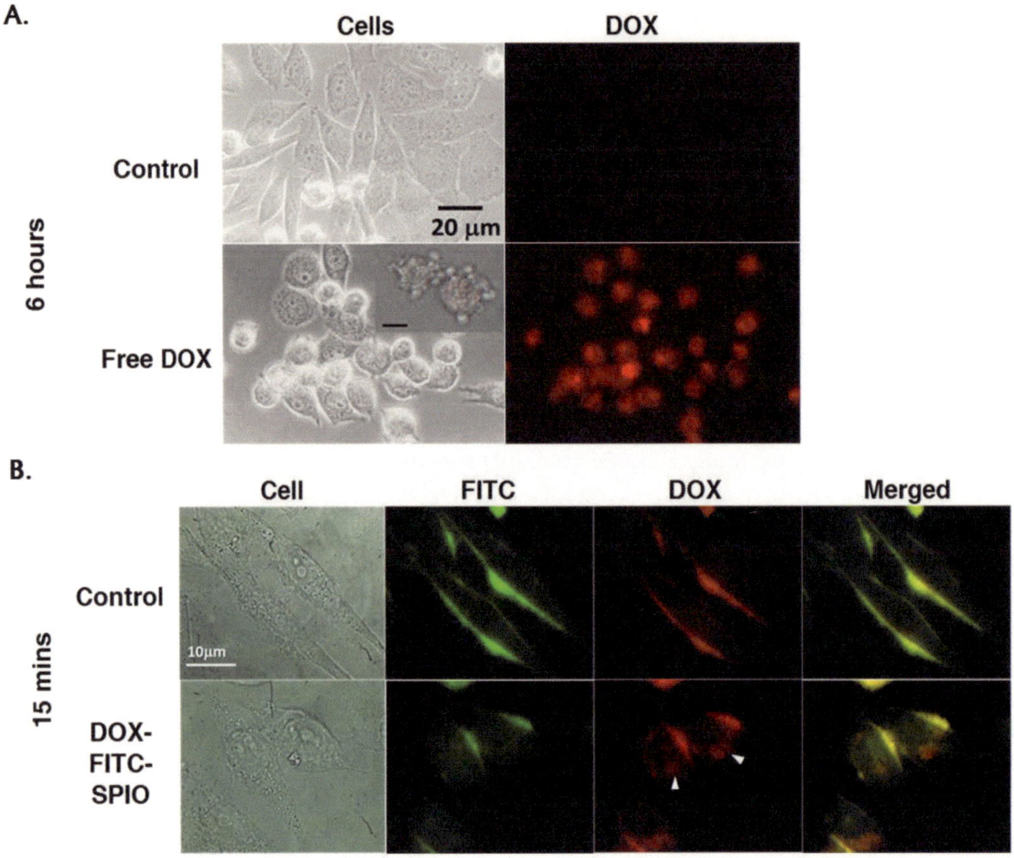

Figure 3. (A) Phase and fluorescent micrograph of MIA PaCa 2 cells incubated with Dox. Note the cellular entry, especially into the nucleus (red fluorescence), of the free drug 6 h following exposure. The inset shows blebbing of cells exposed to Dox prior to cell death (Scale = 20 μm). (B) Phase and fluorescent micrograph of MIA PaCa 2 cells incubated with free Dox and free FITC (control) and Dox-FITC-conjugated Fe_3O_4 NPs at 15 min. Green and red fluorescence represent FITC and Dox, respectively. Note the accumulation of Dox in the nucleus in cells exposed to the Dox and FITC-conjugated SPION. White arrowheads indicate Dox entry into the nucleus. Reprinted from Ref. [105], Copyright (2017), with permission from Elsevier.

5. SPION for cancer diagnosis using MRI

Magnetic resonance imaging with its high spatial resolution has been a preferred method of imaging and diagnosing a disease. It is a noninvasive medical diagnostic tool that monitors the change in magnetization of hydrogen-protons in water molecules contained in a tissue when placed in a magnetic field and exposed to a pulse of radio frequency electromagnetic waves. The mapping of the magnetization provides an image of the organ due to the fact that protons in different tissues, with varying water concentration, respond differently. Contrast agents have been used to enhance the images as they affect the behavior of the protons in their vicinity leading to sharper images. Contrast agents used in MRI are divided into two categories: T_1 and

T_2 contrast agents based on their effect on the magnetic relaxation processes of the protons [106]. Most commonly used T_1 contrast agents are paramagnetic compounds that are composed of metal ions of Gd^{3+} or Mn^{2+} and a chelating ligand, such as diethylene triamine penta-acetic acid, DTPA [106, 107]. The chelate prevents the metal ion from binding to chelates in the body making the paramagnetic ion less toxic. T_1 contrast agents mainly reduce the longitudinal relaxation time (T_1) which is due to energy exchange between the spins and surrounding lattice (spin-lattice relaxation) and result in a brighter signal. T_2 contract agents, consisting of superparamagnetic nanoparticles such as Fe_3O_4, have a strong effect on the transverse relaxation time (T_2). In an external magnetic field, nanoparticles are magnetized and generate induced magnetic fields locally. These induced fields perturb the magnetic relaxation processes of the protons in the water molecules decreasing the T_2 relaxation time, which results in darkening of MR images.

There are various research studies conducted on enhancing the MRI signal for cancer detection using SPIONs as T_2 contrast agents [108–110]. The efficiency of SPIONs as T_2 contrast agents mainly depends on their physicochemical properties, particularly their size and surface chemistry. Stephen et al. [111] report the correlation between particle size and T_2 relaxation. Their study shows that a decrease in particle size leads to reduction in saturation magnetization, which in turn reduces the T_2 relaxation capabilities of SPIONs. There are studies which show the effect of shape on relaxivity. For example, Zhen et al. [112] reported that cubic Fe_3O_4 MNP showed four times smaller relaxation time and thus better image contrast compared to the spherical Fe_3O_4. In another study, octapod Fe_3O_4 nanoparticles with an edge length of 30 nm show a smaller value of T_2 compared to 16-nm spherical Fe_3O_4 nanoparticles possessing a similar M_s [37]. The studies by Park et al. [113] report a decrease in relaxivity as PEG molecular weight increases, indicating that the thickness of PEG coating at the particle surface affects T_2 relaxivity.

When using SPIONs as contrast agents for MRI, it is crucial that they are captured into the cells efficiently upon exposure. Some approaches include introducing peptides [114], antibodies [115], and polymers [116] onto or surrounding magnetic nanoparticles to improve the target specificity. For example, Jun et al. [117] have successfully synthesized superparamagnetic iron oxide nanoparticles of 9-nm size as magnetic probes for the in vivo detection of cancer cells implanted in a mouse. In their research work, 2,3-dimercaptosuccinic acid (DMSA) ligand is attached to the nanoparticles surface to obtain hydrophilic nanoparticles and the nanoparticles are further conjugated with the cancer-targeting antibody, Herceptin. The specific binding properties of Herceptin against a HER2/neu receptor overexpressed from breast cancer cells lead to the successful detection of breast cancer cells (SK-BR-3).

Even though both T_1 and T_2 mapping are powerful techniques, single-mode contrast agents are not always sufficient in modern diagnosis as they have certain drawbacks and limitations [118]. For example, the dark contrast produced by T_2 agents can also be generated from adjacent bones or vascular or there can be susceptibility artifacts due to the sharp change in magnetic field at the surrounding contrast agent. Also, Gd-chelates (T_1 agent) have high mobility which shorten their presence in the vascular system and raise possible toxicity issues. Thus, there is a growing interest in developing complementary T_1-T_2 dual-modal contrast agents, combining the advantages of positive and negative contrasts to obtain high sensitivity and biocompatibility for improved diagnosis [119]. Two different approaches of integrating T_1 and T_2 species have been reported recently [118]. One is constructed by labeling T_1-signaling elements (Gd species) on magnetic nanoparticles. In the study done by Bae et al. [120], Gd-DTPA, a representative Gd chelate-based T_1 MRI contrast agent, is covalently attached to dopamine-coated iron oxide nanoparticles. Their results demonstrated that the composite not only had the ability to improve surrounding water proton signals on the T_1-weighted image but also could induce significant signal reduction on the T_2-weighted image. In another study reported by Santra et al. [121], Gd-DTPA is encapsulated within the poly (acrylic acid) (PAA) polymer-coated SPION (IO-PAA) conjugated to folic acid, which acts as the targeting ligand for breast cancer cells (HeLa cells). When nanoprobes are internalized within the cells, which is acidic, composite magnetic nanoprobe degrades resulting in an intracellular release of Gd-DTPA complex with subsequent T_1 activation, which can be seen by MRI. Authors claim that this T_1 nano-agent could be used for the detection of acidic tumors. The other type of conjugated system consists of T_1 paramagnetic elements embedded into T_2 magnetic nanoparticles. For example, Zhou et al. [122] have synthesized Gd_2O_3-embedded iron oxide nanoparticles with an overall size of 14 nm which can act as a T_1-T_2 mutually enhanced dual-modal contrast agent for MR imaging of liver and hepatic tumor detection with great accuracy in mice. Xiao et al. [123] have prepared PEGylated, Gd-doped iron oxide nanoparticles which is applicable as a T_1-T_2 dual-modal MRI contrast agent. Their in vivo MRI results demonstrated the simultaneous contrast enhancements in T_1- and T_2-weighted MR images toward the glioma-bearing mice.

6. SPION for cancer therapy using magnetic hyperthermia

Magnetic hyperthermia is the transformation of electromagnetic energy from an external alternating magnetic field into heat using MNP. Magnetic nanoparticles serve as the nano-heat centers producing heat by relaxation losses, thereby heating the tissue. The main goal of an effective cancer treatment is to kill the malignant cells with the least of damage to normal cells. As MHT can be used for heating small regions selectively, it offers the potential for being highly selective and noninvasive technique for therapeutic treatment of cancers, and consequently it has advantage over other treatment such as chemotherapy and radiation therapy. It is known that reduced blood flow in tumor causes the lack of oxygen in tumor site which leads to the formation of lactic acid making the cells more acidic [124]. The acidic cells are more sensitive to temperature, have lower thermal resistance than normal cells, and the decreased blood flow in the tumor limit their ability to dissipate heat. As a result, cancer cells can be damaged or killed by increasing the local temperature to the range of 42–46°C with little detriment to healthy cells.

The idea of utilizing SPION for hyperthermia was first proposed by Gilchrist et al. [125] in 1950s and since then many types of MNP are being investigated for this purpose. MNPs have the advantage of being guided and localized specifically at a tumor site by external magnetic fields and can also be directed to the cancer cells by tagging a ligand, such as an antibody or a peptide, without reducing its efficiency. For example, Fabio et al. [126] have reported that the conjugation of folate receptors enhances the targeting for magnetic hyperthermia in solid tumors. Magnetite (Fe_3O_4) and maghemite (γ-Fe_2O_3) have been extensively studied and are promising candidates due to their biocompatibility and relative ease for functionalization. Additionally, iron oxide nanoparticles doped with other magnetic dopants such as Co, Mn, and Ni [127–129] are under investigation to achieve a high heating efficiency by tuning the magnetic anisotropy and saturation magnetization of the material. In addition, many recent findings show that multicore nanoparticles possess a higher heating power than the single-core particles [130], and may offer an advantage. However, among numerous complications, with a high Curie temperature of Fe_3O_4, 850 K, and γ-Fe_2O_3, 750 K, overheating is one of the drawbacks of utilizing these nanoparticles, and as a solution, those complex magnetic oxides with low Curie temperature are being investigated [131–133].

Specific absorption rate (SAR) is a measure of efficiency of heat generation. The SAR value can be estimated by measuring the temperature change in the ferrofluid samples upon exposure to an AC magnetic field following the equation [134]:

$$SAR(T) = \frac{M_{sample}}{m_{Fe_3O_4}} C \left(\frac{\Delta T}{\Delta t}\right)_T \tag{2}$$

Here, M_{sample} is the mass of the sample, $m_{Fe_3O_4}$ is the mass of Fe_3O_4 nanoparticles in the sample, C is the specific heat capacity of the sample, and $\left(\frac{dT}{dt}\right)_T$ is the time rate of change of temperature at T obtained from the slope of the time-dependent temperature data. SAR depends on magnetic properties of the particles such as M_s, anisotropy constant K, particle size distribution (σ), magnetic dipolar interactions, and the rheological properties of the target medium. An ensemble of poly-disperse particles is usually described by a log-normal distribution function,

$$f(D) = \frac{1}{\sqrt{2\pi}\sigma D} \exp\left\{-\frac{[\ln(D/D_\circ)]^2}{2\sigma^2}\right\} \tag{3}$$

Here, D_o is the most probable particle diameter and σ is the width of the distribution. The temperature-dependent average power dissipation in the sample is expressed as [135]

$$\overline{P}(T) = \int_0^\infty \frac{\mu_0 \chi_0 H_0^2 \omega}{2} \frac{\omega \tau_{eff}}{1 + (\omega \tau_{eff})^2} f(D) d(D) \tag{4}$$

where H_o and ω are the amplitude and angular frequency of the applied AC magnetic field, μ_o is the vacuum permeability, τ_{eff} is the effective relaxation time involving Néel relaxation and the Brownian relaxation times, and χ_o is equilibrium susceptibility. τ_{eff} is defined as $\frac{1}{\tau_{eff}} = \frac{1}{\tau_N} + \frac{1}{\tau_B}$, where $\tau_B = \frac{4\pi\eta R_H^3}{k_B T}$ and $\tau_N = \frac{\sqrt{\pi}}{2}\tau_0 \exp\left(\frac{KV}{k_B T}\right)\sqrt{\frac{KV}{k_B T}}$ are the Néel and Brownian relaxation times, η is the viscosity of the suspension, R_H is the hydrodynamic radius of the coated nanoparticle, V_m is the magnetic volume of the nanoparticles, and $\tau_o \sim 10^{-9}$ s. χ_o is given by $\chi_o = \chi_i\left(\coth\xi - \frac{1}{\xi}\right)$, where $\chi_i = \frac{\mu_\circ \phi M_d^2 H_\circ V_m}{3k_b T}$ is the initial susceptibility, and $\xi = \frac{\mu_\circ M_d H_\circ V_m}{k_b T}$, with M_d being the domain magnetization of the nanoparticle and ϕ the volume fraction of the magnetic nanoparticles in the ferrofluid. SAR in units of W/g is obtained using Eq. (4) as $\overline{P}(T)/m_{Fe_3O_4}$, where $m_{Fe_3O_4}$ is the mass of Fe_3O_4 nanoparticles in ferrofluids.

MHT investigations are often done on colloidal suspensions of surface-coated MNP, called ferrofluids. It is often necessary to coat the MNP using a biocompatible polymer to avoid direct contact with the tissue and to reduce the particle aggregation. Ferrofluid preparations frequently yield a mixture of isolated nanoparticles and nanoclusters [136] with varying degree of magnetic dipole-dipole interactions present in ferrofluids. It has been shown that the dipolar interactions among the MNP affect the SAR value drastically and can be exploited to optimize SAR [136–139]. A mean-field approximation method has been used to account for effects of interactions on SAR for a collection of monodisperse MNP [138, 139], and we have used this approach to explain the very different observed SAR values for similar size particles prepared by two different methods of preparation [140].

Since the surfactant influences the magnetic properties as well as the degree of interactions in the ferrofluid, a useful approach for improving the magnetic hyperthermia performance is to optimize the surface coating to maximize the SAR. Currently, there are conflicting SAR values obtained for a certain sized nanoparticle making it difficult to evaluate the exact contributions of surface coating on the SAR. According to Mohammad et al. [141], it is found that inorganic coatings improve the SAR value and the gold coating retains the superparamagnetic fraction of Fe_3O_4 nanoparticles much better than uncoated nanoparticles alone and leads to higher magnetocrystalline anisotropy. A study by Liu et al. [142] suggested the possibility of increasing SAR by decreasing the surface-coating thickness using highly monodispersed Fe_3O_4 nanoparticles with different polyethylene glycol-coating thickness. The increase in SAR was explained as due to a decrease in coating thickness leading to an increased Brownian loss, improved thermal conductivity, as well as improved dispersion. It should be noted that the heating performance of the nanoparticles depends on the medium as well. Whenever nanoparticles encounter biological systems, interactions take place between their surfaces and biological components such as proteins, membranes, phospholipids, and DNA forming the so-called protein corona around the nanoparticles [143, 144]. The formation of corona depends on the surface properties of the particles [145, 146] and can influence the aggregation behavior of nanoparticles in biological media, which in turn can affect their performance for desired applications. Therefore, apart from the optimization of the properties of the magnetic core and surface coating for high-performance MHT, it is necessary to ensure its performance in the physiological environments.

In the work reported by Khandhar et al. [147], authors use poly(maleic anhydride-alt-1-octadecene)-poly(ethylene glycol) (PMAO-PEG), an amphiphilic polymer-coated Fe_3O_4 nanoparticles of three different sizes, 13, 14, and 16 nm, to study the MHT efficiency in cell growth medium (CGM) similar to biological environment. Their results showed an increase in hydrodynamic sizes in all three samples upon exposure to CGM. SAR reduced (30%) only in 16-nm size sample, while other two samples did not exhibit any significant decrease in SAR. The authors suggest that the increase in hydrodynamic volume prolongs Brownian relaxation while Néel relaxation is unaffected. Hence, in 13- and 14-nm samples where SAR is mainly due to Néel relaxation, SAR was not affected. But in the 16-nm samples, in which there is a contribution from Brownian relaxation to heat dissipation, the SAR dropped due to the increase in Brownian relaxation. We have investigated the magnetic hyperthermia efficiency of dextran and citric acid (CA)-coated Fe_3O_4 ferrofluids in cell growth medium, which contains serum proteins similar to physiological environments. From the stock solutions (25 mg/ml), 3 mg/ml concentration of dextran and citric acid-coated ferrofluids samples were prepared using CGM and deionized water (DI) water. The ferrofluid samples were subjected to an AC field of 235-Oe amplitude at the frequency of 375 kHz. The SAR of dextran-coated samples in DI and CGM was estimated to be 63 and 72 W/g, which indicates that their performance is not much affected by the medium if not enhanced. However, SAR values obtained for CA-coated samples in DI and CGM, 78 and 38 W/g, implies that their efficiency is heavily reduced when exposed to physiological environments.

7. Gd-doped SPION as a potential theranostic agent

The multifunctionality of SPION makes them a good candidate for theranostics. One such approach to integrate diagnostic imaging and therapeutic function is to develop SPION as an MRI/a drug delivery platform. Yu et al. [148] reported that PEG-coated iron oxide nanoparticles when loaded with Dox provide a therapeutic capability. Following their injection into a mouse, Dox-modified magnetic nanoparticles accumulate in the tumor and the nanoparticles were imaged by using T_2 MRI. The contrast associated with the tumor changes from light to dark at 4.5 h post injection and the growth rate of the tumor mass was decreased in the nanoparticle-injected mice compared to that of a control group. In another work, Lee et al. [149] developed PEG-stabilized Fe_3O_4 nanocrystals on dye-doped mesoporous silica nanoparticles and Dox was loaded into the pores. Here, SPIONs work as a contrast agent in MRI, the dye molecule imparts optical imaging modality, and Dox induces cell death. In a similar approach, Kim et al. [150] developed a core-shell structure consisting of single Fe_3O_4 core and mesoporous silica shell for MR and fluorescence imaging which also has the potential to be used as a drug carrier. Hayashi et al. [92] reports a study done on SPION conjugated with folic acid as targeting ligand and Tamoxifen as anticancer drug. The drug release was triggered by heat generated by SPION in an AC magnetic field, hence performing drug delivery and hyperthermia simultaneously.

The incorporation of MHT and imaging modalities [151] has been investigated widely as well. One such study is reported by Hayashi et al. [152] in which authors have investigated SPION for cancer theranostics by combining MRI and magnetic hyperthermia through a set of *in vivo* experiments. They show that FA- and PEG-modified SPION nanoclusters accumulated locally in cancer tissues within the tumor and enhanced the MRI contrast. Furthermore, they report that with MHT, the tumor volume of treated mice was reduced to one-tenth that of the control mice. Also, Gd-doped Fe_3O_4 nanoparticles have the potential to act as an effective MHT agent [153] in addition to their use as a T_1-T_2 dual-modal contrast agent for MR imaging. Gd(III) is known to oppose net magnetic moment of Fe(III)/Fe(II) in oxides, reducing magnetization [154–156]. Therefore, Gd doping may reduce the hyperthermia efficiency, but by using the correct amount of doping, one can always explore the possibility of using them as both MRI contrast agents and hyperthermia mediators. However, there are very few studies done on the MHT efficiency of Gd-doped Fe_3O_4 nanoparticles [157, 158]. Both the studies report higher SAR values for Gd-doped Fe_3O_4 nanoparticles compared to the reported values for undoped samples. In our work presented here, we have investigated the MHT efficiency of $Gd_{0.075}Fe_{2.925}O_4$ nanoparticles for possible use as a theranostic agent.

7.1. Synthesis and characterization of Gd-doped SPION

$Gd_{0.075}Fe_{2.925}O_4$ nanoparticles were synthesized by coprecipitation method. For a typical synthesis of $Gd_{0.075}Fe_{2.925}O_4$, aqueous solution of $FeCl_2.4H_2O$, $FeCl_3.6H_2O$ and $Gd(NO_3)_3$ was mixed in a molar ratio of 1.00:1.925:0.075 in 25-ml volume followed by the addition of 250 ml of 1 M NH_4OH. The synthesized nanoparticles were then coated with dextran according to the method outlined by Arachchige et al. [140]. From the structural investigation, it was observed that Gd doping does not alter the Fe_3O_4 crystal structure significantly (**Figure 4**). Using several intense X-ray diffraction (XRD) peaks and the Debye-Scherer equation, the crystallite sizes of the Fe_3O_4 and $Gd_{0.075}Fe_{2.925}O_4$ nanoparticle samples were determined to be 11.7 ± 0.6 and 14.9 ± 0.5 nm, respectively. This increase in the crystallite size is consistent with the previous studies on Gd doping in spinel structures [159].

TEM images of the two samples are shown in **Figure 5**. The undoped sample consists of roughly spherical nanoparticles with smaller polydispersity, whereas the Gd-doped sample exhibits nanoparticles with rough edges with wider size distribution.

The magnetic properties of the synthesized powder as well as the ferrofluid samples are determined by analyzing the *M*(*H*) curve. The M-H data for undoped and Gd-doped Fe_3O_4 ferrofluid samples, recorded at room temperature, are shown in **Figure 6**. The sigmoidal shape of the *M*(*H*) curves with nearly zero hysteresis confirms the superparamagnetic nature of these nanoparticles at room temperature. The saturation magnetization of Fe_3O_4 nanoparticles is measured to be ~72 emu/g, whereas that of Gd-doped Fe_3O_4 nanoparticles is reduced to ~52 emu/g. This reduction in saturation magnetization at room temperature agrees with the observations in other reported studies [123, 158] and can be attributed to the fact that magnetic Fe^{3+} ions get replaced by the Gd^{3+} ions in the octahedral sites of the inverse spinel structure.

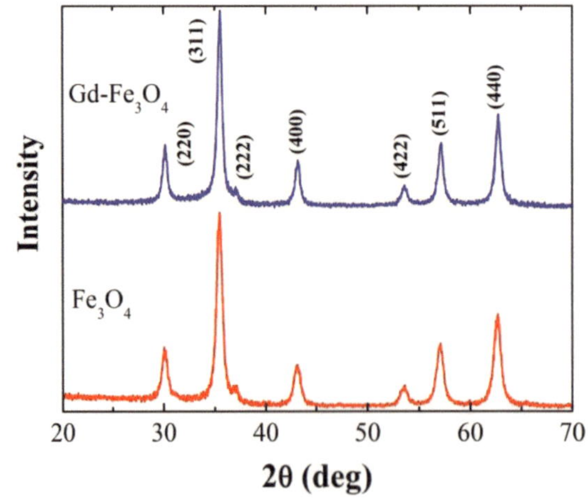

Figure 4. X-ray diffraction patterns of as-prepared Fe_3O_4 and $Gd_{0.075}Fe_{2.925}O_4$ nanoparticles [160].

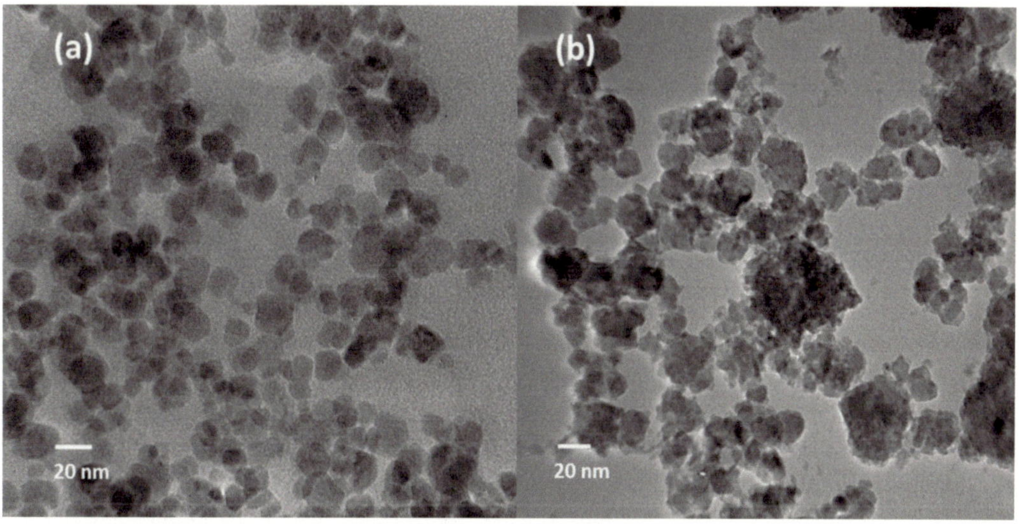

Figure 5. TEM images of (a) Fe_3O_4 and (b) Gd-Fe_3O_4 nanoparticles [160].

It is observed that the doping of Gd^{3+} ions into Fe_3O_4 spinel has significantly influenced the average crystallite size and the saturation magnetization. The *M*(*H*) curve for an ensemble of noninteracting superparamagnetic nanoparticles described by a log-normal distribution function, *f*(*D*), can be fitted using the following expression:

$$M(H) = M_s \frac{\int_0^\infty f(D)VL(x)dD}{\int_0^\infty f(D)VdD} \tag{5}$$

where $L(x) = \coth x - \frac{1}{x}$ is the Langevin function, $x = (M_s VH)/k_B T$, M_s is the saturation magnetization, and *V* is the volume of the particle. The fitted particle size was inconsistent with the

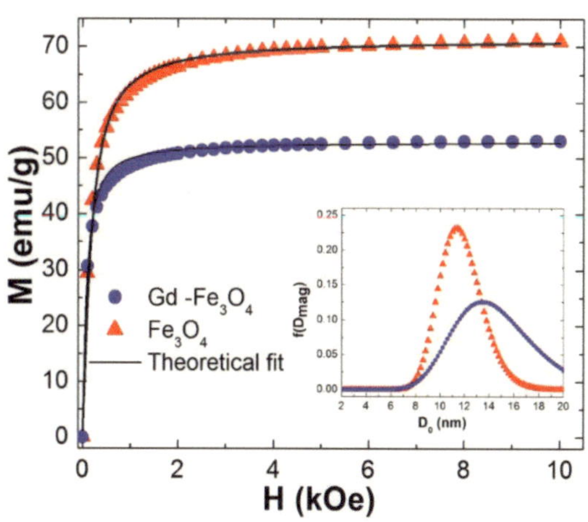

Figure 6. *M* versus *H* curves for two ferrofluid samples fitted with Eq. (5). The inset shows the resulting particle size distribution obtained for the two samples [160].

observed XRD data, and it was necessary to introduce the magnetic dipolar interaction effects through a phenomenological temperature, T^*, as described in our recent work [140]. The best-fit parameters for the two samples are shown in **Table 2**, and **Figure 6** shows the fitted data. The inset in **Figure 6** shows the magnetic core size distributions for two ferrofluid samples. The fitting of *M*(*H*) data with Eq. (5) clearly shows that Gd-doped Fe_3O_4 nanoparticles have a higher average magnetic core size with a larger size distribution (14.6 ± 3.7 nm) and lower saturation magnetization (52 emu/g) compared to the undoped Fe_3O_4 nanoparticles (11.7 ± 1.9 nm, 72 emu/g). Both the ferrofluid samples exhibit similar strength of magnetic dipolar interaction (T^* ~80–100 K).

MHT measurements were carried out on the dextran-coated Gd-doped as well as undoped Fe_3O_4 ferrofluid samples at a field of 235 Oe and at a frequency of 375 kHz. The heating curves for two samples are shown in **Figure 7(a)**, and from the plot it can be observed that the initial heating rates for the two samples are approximately the same. From these heating curves, the SAR values were obtained as a function of temperature taking into account heat loss as described elsewhere [134]. **Figure 7(b)** presents the corrected experimental SAR data as a function of temperature for both the undoped and Gd-doped samples. Within the experimental error, the room temperature SAR values for Gd-doped Fe_3O_4 and undoped

Ferrofluid sample	M_s (emu/g)	D_o (nm)	σ	T^* (K)	D_{avg} (nm)
Fe_3O_4	72	11.6	0.15	80	11.7 ± 1.9
Gd -Fe_3O_4	52	14.2	0.23	100	14.6 ± 3.7

Table 2. Fitting parameters obtained from the *M*(*H*) fitting with modified Langevin function using T^*.

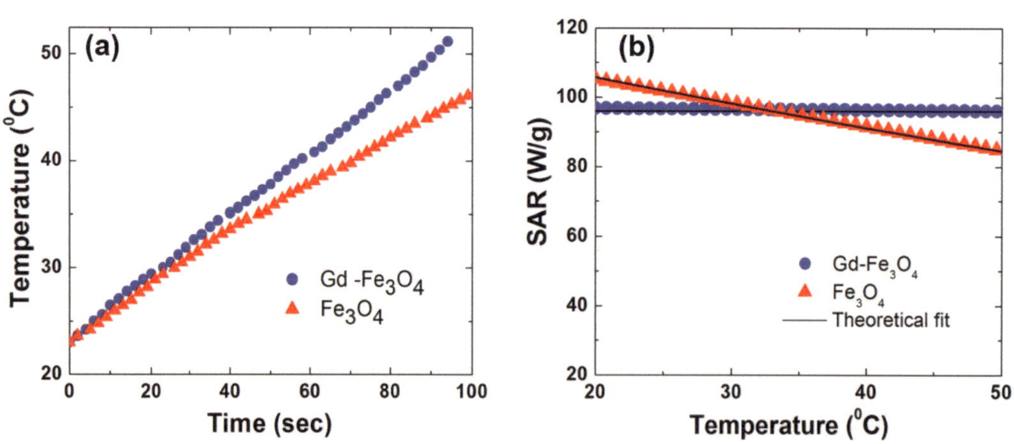

Figure 7. (a) Heating profiles of Fe_3O_4 and Gd-doped Fe_3O_4 ferrofluid samples under an AC magnetic field amplitude of 235 Oe and at a frequency of 375 kHz. (b) The temperature dependence of net SAR for two ferrofluid samples. The black line shows the theoretical fitting of the experimental data with the linear response theory [160].

Fe_3O_4 ferrofluid are very similar. The temperature-dependent SAR values were fitted to the linear response theory incorporated with the interactions and size distribution [140]. The solid lines in **Figure 7(b)** are the best fits to the experimental SAR data, using the particle size distribution parameters and T^* values given in **Table 2** and treating the anisotropy constant, K, as a fitting parameter. The SAR fitting yields a somewhat smaller anisotropy constant (~12 kJ/m^3) for Gd-doped ferrofluid compared to that of undoped ferrofluid (~21kJ/m^3). It is interesting to note that both samples have similar SAR values in spite of a smaller anisotropy constant and the saturation magnetization for the $Gd_{0.075}Fe_{2.925}O_4$ sample compared to the undoped sample. The expected lowering of SAR in $Gd_{0.075}Fe_{2.925}O_4$ is probably offset by a larger particle size in this sample, as the SAR would increase with increasing particle size up to a critical size [140]. By fine-tuning the composition of Gd-doped Fe_3O_4 nanoparticles, we may achieve a higher SAR value.

In summary, the Gd doping on the Fe_3O_4 nanoparticles affects the morphology and the magnetic properties of Fe_3O_4 nanoparticles considerably but the magnetic hyperthermia efficiency of the samples was about the same within the experimental uncertainties. The possibility of using Gd-doped Fe_3O_4 nanoparticles as a dual-modal T_1-T_2 contrast agent is being currently explored by others and our magnetic hyperthermia results demonstrate that this material is a potential candidate for multimodal contrast imaging and cancer treatment by hyperthermia. However, further research is necessary to optimize the amount of Gd doping to enhance SAR for cancer treatment and to be used as a theranostic agent.

8. Conclusions

In this chapter, we have discussed various approaches to exploit the multifunctionality of SPION for cancer theranostics. We have given a brief background on the nanoparticle magnetism, followed by a description of commonly used synthesis methods and surface-functionalizing strategies. Three major applications of Fe_3O_4 nanoparticles in drug delivery, MRI, and MHT, including our recent work on Gd-doped SPION as a possible theranostic agent, are described. This chapter also addresses the recent work on integrating the individual diagnostic and therapeutic approaches to develop SPION-based theranostic platform.

Despite the exciting progress, SPION is far from meeting clinical standards as theranostic agent. It has its own promises and advantages, but there are still some disadvantages to be overcome. These include target specificity as drug carriers, optimizing the heating efficiency and aim for sufficient heating using minimum dosage, preventing the overheating in MHT, and issues of biocirculation, biodistribution, and bioelimination within the biological system. In summary, although in theory, SPION is a perfect vehicle in the development of theranostic nanomedicine, more research is required to overcome its disadvantages and this should be the main focus of the next stages of investigation.

9. Future directions

In recent years, the research in the field of theranostics has brought many diverse fields together for targeting, imaging, and therapy for a deadly disease like cancer. These fields include physics of magnetism, chemistry of synthesis, material science of structure-property relationship, surface science for functionalization, biomedical engineering in MRI and radiofrequency activation and treatment, and biology for understanding the behavior of cancer cells. SPIONs have played a key role in this application as visual, imaging, and therapeutic agent. Several studies have shown promising results; however, many challenges still remain in moving theranostic applications from laboratory settings to clinics. Two major challenges we face are low efficacy and toxicity of SPION. For in vivo applications, the amount of SPION used (several hundred microgram/ml) usually produces undesirable toxic side effects. The smaller concentration, on the other hand, is not sufficient for imagining and therapeutic action of the material. It is well known that the size, shape, and surface modifications influence the performance of SPION. There is a lack of information about the combined effects of these parameters in the clinical applications. Also, we do not have a clear understanding of controlling the delivery of SPION to a specific target in the body by using external magnetic field gradient. It has been found that some particles would end up accumulating in other parts of the body such as liver, spleen, kidney, and lungs along with the specific intended location. We do not know how they will affect those nonspecific organs and how long they will stay there. In magnetic hyperthermia therapy, measuring precise temperature at the tumor site and adjusting particle properties with frequencies and amplitude of the external field for apoptosis/necrosis of cancer cells without affecting normal tissues are challenges that researchers and clinicians face every day. There have been promising results in treating prostate and skin cancers with magnetic fluid hyperthermia but no real efforts have been made to treat deep tumors such a pancreas and liver with SPION. In order to make progress with these therapies, research is needed in the development of new materials that have higher reflexivity, better thermal activation properties, and have better coating materials to improve the bio-distribution and biocompatibility for in vivo applications. Most imperatively, we need data on large animal studies before theranostics can make a fruitful transition from research laboratories to the clinics, and so on.

Acknowledgements

RN would like to thank Richard Barber Foundation for continued research support for this work. PPV would like to thank Kettering University for financial support through multiple Faculty Research Fellowship and Rodes professorship awards.

http://dx.doi.org/10.5772/intechopen.68219

Author details

Maheshika Palihawadana-Arachchige[1], Vaman M. Naik[2], Prem P. Vaishnava[3], Bhanu P. Jena[4] and Ratna Naik[1]*

*Address all correspondence to: rnaik@wayne.edu

1 Department of Physics and Astronomy, Wayne State University, Detroit, MI, USA

2 Department of Natural Sciences, University of Michigan-Dearborn, Dearborn, MI, USA

3 Kettering University, Flint, MI, USA

4 Department of Physiology, State University, Detroit, MI, USA

References

[1] Portney NG, Ozkan M. Nano-oncology: Drug delivery, imaging, and sensing. Analytical and Bioanalytical Chemistry. 2006;**384**:620–630

[2] Farokhzad OC, Langer R. Nanomedicine: Developing smarter therapeutic and diagnostic modalities. Advanced Drug Delivery Reviews. 2006;**58**:1456–1459

[3] Peer D, Karp JM, Hong S, Farokhzad OC, Margalit R, Langer R. Nanocarriers as an emerging platform for cancer therapy. Nature Nanotechnology. 2007;**2**:751–760

[4] Yoo D, Lee J-H, Shin T-H, Cheon J. Theranostic magnetic nanoparticles. Accounts of Chemical Research. 2011;**44**:863–874

[5] Xie J, Lee S, Chen X. Nanoparticle-based theranostic agents. Advanced Drug Delivery Reviews. 2010;**62**:1064–1079

[6] Veiseh O, Gunn JW, Zhang M. Design and fabrication of magnetic nanoparticles for targeted drug delivery and imaging. Advanced Drug Delivery Reviews. 2010;**62**:284–304

[7] Rosen JE, Chan L, Shieh D-B, Gu FX. Iron oxide nanoparticles for targeted cancer imaging and diagnostics. Nanomedicine Nanotechnology. 2012;**8**:275–290

[8] Xie J, Jon S. Magnetic nanoparticle-based theranostics. Theranostics. 2012;**2**:122–124

[9] Ito A, Shinkai M, Honda H, Kobayashi T. Medical application of functionalized magnetic nanoparticles. Journal of Bioscience and Bioengineering.2005;**100**:1–11

[10] Colombo M, Carregal-Romero S, Casula MF, Gutiérrez L, Morales MP, Böhm IB, Heverhagen JT, Prosperi D, Parak WJ. Biological applications of magnetic nanoparticles. Chemical Society Review. 2012;**41**:4306–4334

[11] Issa B, Obaidat IM, Albiss BA, Haik Y. Magnetic nanoparticles: Surface effects and properties related to biomedicine applications. International Journal of Molecular Science. 2013;**14**:21266–21305

[12] Huang S-H, Juang R-S. Biochemical and biomedical applications of multifunctional magnetic nanoparticles: A review. Journal of Nanoparticle Research. 2011;**13**:4411

[13] Lawes G, Naik R, Vaishnava P. Physical properties and biomedical applications of superparamagnetic iron oxide nanoparticles. Nanocellbiology. 2014:257

[14] Kolhatkar AG, Jamison AC, Litvinov D, Willson RC, Lee TR. Tuning the magnetic properties of nanoparticles. International Journal of Molecular Science. 2013;**14**:15977–16009

[15] Guardia P, Labarta A, Batlle X. Tuning the size, the shape, and the magnetic properties of iron oxide nanoparticles. Journal of Physical Chemistry C. 2011;**115**:390–396

[16] Singamaneni S, Bliznyuk VN, Binek C, Tsymbal EY. Magnetic nanoparticles: Recent advances in synthesis, self-assembly and applications. Journal of Material Chemistry. 2011;**21**:16819–16845

[17] Lu AH, Salabas EEL, Schüth F. Magnetic nanoparticles: Synthesis, protection, functionalization, and application. Angewandte Chemie International Edition. 2007;**46**:1222–1244

[18] Gupta AK, Gupta M. Synthesis and surface engineering of iron oxide nanoparticles for biomedical applications. Biomaterials. 2005;**26**:3995–4021

[19] Pisane KL, Singh S, Seehra MS. Synthesis, structural characterization and magnetic properties of Fe/Pt core-shell nanoparticles. Journal of Applied Physics. 2015;**117**:17d708

[20] Singh V, Seehra MS, Bali S, Eyring EM, Shah N, Huggins FE, Huffman GP. Magnetic properties of (Fe, Fe–B)/γ-Fe_2O_3 core shell nanostructure. Journal of Physics and Chemistry of Solids. 2011;**72**:1373–1376

[21] Seehra MS, Singh V, Dutta P, Neeleshwar S, Chen YY, Chen CL, Chou SW, Chen CC. Size-dependent magnetic parameters of fcc FePt nanoparticles: Applications to magnetic hyperthermia. Journal of Physics D: Applied Physics. 2010;**43**:145002

[22] Pisane KL, Despeaux EC, Seehra MS. Magnetic relaxation and correlating effective magnetic moment with particle size distribution in maghemite nanoparticles. Journal of Magnetism and Magnetic Materials. 2015;**384**:148–154

[23] Seehra MS, Pisane KL. Relationship between blocking temperature and strength of interparticle interaction in magnetic nanoparticle systems. Journal of Physics and Chemistry of Solids. 2016;**93**:79–81

[24] Pankhurst QA, Connolly J, Jones SK, Dobson J. Applications of magnetic nanoparticles in biomedicine. Journal of Physics D: Applied Physics. 2003;**36**:R167

[25] Jun Y-W, Seo J-W, Cheon J. Nanoscaling laws of magnetic nanoparticles and their applicabilities in biomedical sciences. Accounts of Chemical Research. 2008;**41**:179–189

[26] Guardia P, Pérez N, Labarta A, Batlle X. Controlled synthesis of iron oxide nanoparticles over a wide size range. Langmuir. 2010;**26**:5843–5847

[27] Comesaña-Hermo M, Ciuculescu D, Li Z-A, Stienen S, Spasova M, Farle M, Amiens C. Stable single domain Co nanodisks: Synthesis, structure and magnetism. Journal of Material Chemistry. 2012;**22**:8043–8047

[28] Cozzoli PD, Snoeck E, Garcia MA, Giannini C, Guagliardi A, Cervellino A, Gozzo F, Hernando A, Achterhold K, Ciobanu N, Parak FG, Cingolani R, Manna L. Colloidal synthesis and characterization of tetrapod-shaped magnetic nanocrystals. Nano Letters. 2006;**6**:1966–1972

[29] Gao G, Liu X, Shi R, Zhou K, Shi Y, Ma R, Takayama-Muromachi E, Qiu G. Shape-controlled synthesis and magnetic properties of monodisperse Fe_3O_4 nanocubes. Crystal Growth & Design. 2010;**10**:2888–2894

[30] Han GC, Zong BY, Wu YH. Magnetic properties of magnetic nanowire arrays. IEEE Transactions on Magnetics. 2002;**38**:2562–2564

[31] Puntes VF, Zanchet D, Erdonmez CK, Alivisatos AP. Synthesis of hcp-Co nanodisks. Journal of the American Chemical Society. 2002;**124**:12874–12880

[32] Schladt TD, Shukoor MI, Schneider K, Tahir MN, Natalio F, Ament I, Becker J, Jochum FD, Weber S, Köhler O, Theato P, Schreiber LM, Sönnichsen C, Schröder HC, Müller WEG, Tremel W. Au@MnO nanoflowers: Hybrid nanocomposites for selective dual functionalization and imaging. Angewandte Chemie International Edition. 2010;**49**: 3976–3980

[33] Song Q, Zhang ZJ. Shape control and associated magnetic properties of spinel cobalt ferrite nanocrystals. Journal of the American Chemical Society. 2004;**126**:6164–6168

[34] Wu C-G, Lin HL, Shau N-L. Magnetic nanowires via template electrodeposition. Journal of Solid State Electrochemistry. 2006;**10**:198–202

[35] Yan M, Fresnais J, Berret J-F. Growth mechanism of nanostructured superparamagnetic rods obtained by electrostatic co-assembly. Soft Matter. 2010;**6**:1997–2005

[36] Závišová V, Tomašovičová N, Kováč J, Koneracká M, Kopčanský P, Vávra I. Synthesis and characterisation of rod-like magnetic nanoparticles. Olomouc, Czech Republic. 2010;**10**.

[37] Zhao Z, Zhou Z, Bao J, Wang Z, Hu J, Chi X, Ni K, Wang R, Chen X, Chen Z, Gao J. Octapod iron oxide nanoparticles as high-performance T2 contrast agents for magnetic resonance imaging. Nature Communications. 2013;**4**:2266

[38] Salazar-Alvarez G, Qin J, Sepelak V, Bergmann I, Vasilakaki M, Trohidou K, Ardisson J, Macedo W, Mikhaylova M, Muhammed M. Cubic versus spherical magnetic nanoparticles: The role of surface anisotropy. Journal of the American Chemistry Society. 2008;**130**:13234–13239

[39] Noh S-H, Na W, Jang J-T, Lee J-H, Lee EJ, Moon SH, Lim Y, Shin J-S., Cheon J. Nanoscale magnetism control via surface and exchange anisotropy for optimized ferrimagnetic hysteresis. Nano Letters. 2012;**12**:3716–3721

[40] Lee J-H, Huh Y-M, Jun Y-W, Seo J-W, Jang J-T, Song H-T, Kim S, Cho E-J, Yoon H-G, Suh J-S, Cheon J. Artificially engineered magnetic nanoparticles for ultra-sensitive molecular imaging. Nature Medicine. 2007;**13**:95–99

[41] Grasset F, Mornet S, Demourgues A, Portier J, Bonnet J, Vekris A, Duguet E. Synthesis, magnetic properties, surface modification and cytotoxicity evaluation of $Y_3Fe_{5-x}Al_xO_{12}$ ($0 \leqslant x \leqslant 2$) garnet submicron particles for biomedical applications. Journal of Magnetism and Magnetic Materials. 2001;**234**:409–418

[42] Sun S, Zeng H. Size-controlled synthesis of magnetite nanoparticles. Journal of the American Chemistry Society. 2002;**124**:8204–8205

[43] Jana NR, Chen Y, Peng X. Size- and shape-controlled magnetic (Cr, Mn, Fe, Co, Ni) oxide nanocrystals via a simple and general approach. Chemistry of Materials. 2004;**16**:3931–3935

[44] Hyeon T. Chemical synthesis of magnetic nanoparticles. Chemical Communications. 2003:927–934

[45] Roca AG, Costo R, Rebolledo AF, Veintemillas-Verdaguer S, Tartaj P, González-Carreño T, Morales MP, Serna CJ. Progress in the preparation of magnetic nanoparticles for applications in biomedicine. Journal of Physics D: Applied Physics. 2009;**42**:224002

[46] Laurent S, Forge D, Port M, Roch A, Robic C, Vander Elst L, Muller R.N. Magnetic iron oxide nanoparticles: Synthesis, stabilization, vectorization, physicochemical characterizations, and biological applications. Chemical Reviews. 2008;**108**:2064–2110

[47] Mascolo M, Pei Y, Ring T. Room temperature co-precipitation synthesis of magnetite nanoparticles in a large pH window with different bases. Materials. 2013;**6**:5549

[48] Latham AH, Williams ME. Controlling transport and chemical functionality of magnetic nanoparticles. Accounts of Chemical Research. 2008;**41**:411–420

[49] Daou TJ, Pourroy G, Bégin-Colin S, Grenèche JM, Ulhaq-Bouillet C, Legaré P, Bernhardt P, Leuvrey C, Rogez G. Chemistry of materials, hydrothermal synthesis of monodisperse magnetite nanoparticles. American Chemical Society. 2006:4399–4404

[50] Ge S, Shi X, Sun K, Li C, Uher C, Baker JR, Banaszak Holl MM, Orr BG. Facile hydrothermal synthesis of iron oxide nanoparticles with tunable magnetic properties. Journal of Physics and Chemistry C. 2009;**113**:13593–13599

[51] Chin AB, Yaacob II. Synthesis and characterization of magnetic iron oxide nanoparticles via w/o microemulsion and Massart's procedure. Journal of Materials Processing Technology. 2007;**191**:235–237

[52] Wei W, Quanguo H, Hong C, Jianxin T, Libo N. Sonochemical synthesis, structure and magnetic properties of air-stable Fe_3O_4/Au nanoparticles. Nanotechnology. 2007;**18**: 145609

[53] Alexis F, Pridgen E, Molnar LK, Farokhzad OC. Factors affecting the clearance and biodistribution of polymeric nanoparticles. Molecular Pharmaceutics. 2008;**5**:505–515

[54] Knop K, Hoogenboom R, Fischer D, Schubert US. Poly(ethylene glycol) in drug delivery: Pros and cons as well as potential alternatives. Angewandte Chemie International Edition. 2010;**49**:6288–6308

[55] Filippousi M, Angelakeris M, Katsikini M, Paloura E, Efthimiopoulos I, Wang Y, Zamboulis D, Van Tendeloo G. Surfactant effects on the structural and magnetic properties of iron oxide nanoparticles. Journal of Physics and Chemistry C. 2014;**118**:16209–16217

[56] Marín T, Montoya P, Arnache O, Calderón J. Influence of surface treatment on magnetic properties of Fe_3O_4 nanoparticles synthesized by electrochemical method. Journal of Physics and Chemistry B. 2016

[57] Soares PIP, Alves AMR, Pereira LCJ, Coutinho JT, Ferreira IMM, Novo CMM, Borges JPMR. Effects of surfactants on the magnetic properties of iron oxide colloids. Journal of Colloid and Interface Science. 2014;**419**:46–51

[58] Yuan Y, Rende D, Altan CL, Bucak S, Ozisik R, Borca-Tasciuc D-A. Effect of surface modification on magnetization of iron oxide nanoparticle colloids. Langmuir. 2012;**28**: 13051–13059

[59] Salafranca J, Gazquez J, Pérez N, Labarta A, Pantelides ST, Pennycook SJ, Batlle X, Varela M. Surfactant organic molecules restore magnetism in metal-oxide nanoparticle surfaces. Nano Letters. 2012;**122**:499–2503

[60] Duan H, Kuang M, Wang X, Wang YA, Mao H, Nie S. Reexamining the effects of particle size and surface chemistry on the magnetic properties of iron oxide nanocrystals: New insights into spin disorder and proton relaxivity. Journal of Physics and Chemistry C. 2008;**112**:8127–8131

[61] Molday RS, MacKenzie D. Immunospecific ferromagnetic iron-dextran reagents for the labeling and magnetic separation of cells. Journal of Immunological Methods. 1982;**52**: 353–367

[62] Bulte JWM, Ma LD, Magin RL, Kamman RL, Hulstaert CE, Go KG, The TH, De Leij L. Selective MR imaging of labeled human peripheral blood mononuclear cells by liposome mediated incorporation of dextran-magnetite particles. Magnetic Resonance in Medicine. 1993;**29**:32–37

[63] Massia SP, Stark J, Letbetter DS. Surface-immobilized dextran limits cell adhesion and spreading. Biomaterials. 2000;**21**:2253–2261

[64] Weissleder R, Elizondo G, Wittenberg J, Lee AS, Josephson L, Brady TJ. Ultrasmall superparamagnetic iron oxide: An intravenous contrast agent for assessing lymph nodes with MR imaging. Radiology. 1990;**175**:494–498

[65] Josephson L, Tung C-H, Moore A, Weissleder R. High-efficiency intracellular magnetic labeling with novel superparamagnetic-tat peptide conjugates. Bioconjugate Chemistry. 1999;**10**:186–191

[66] Wunderbaldinger P, Josephson L, Weissleder R. Crosslinked iron oxides (CLIO). Academic Radiology. 2002;**9**:S304-S306

[67] Mikhaylova M, Kim DK, Bobrysheva N, Osmolowsky M, Semenov V, Tsakalakos T, Muhammed M. Superparamagnetism of magnetite nanoparticles: Dependence on surface modification. Langmuir. 2044;**20**:2472–2477

[68] Kim DK, Mikhaylova M, Wang FH, Kehr J, Bjelke B, Zhang Y, Tsakalakos T, Muhammed M. Starch-coated superparamagnetic nanoparticles as MR contrast agents. Chemistry of Materials. 2003;**15**:4343–4351

[69] Donadel K, Felisberto MDV, Fávere VT, Rigoni M, Batistela NJ, Laranjeira MCM. Synthesis and characterization of the iron oxide magnetic particles coated with chitosan biopolymer. Material Science Engineering: C. 2008;**28**:509–514

[70] Denkbaş EB, Kiliçay E, Birlikseven C, Öztürk E. Magnetic chitosan microspheres: Preparation and characterization. Reactive and Functional Polymers. 2002;**50**:225–232

[71] Kim EH, Ahn Y, Lee HS. Biomedical applications of superparamagnetic iron oxide nanoparticles encapsulated within chitosan. Journal of Alloys Compounds. 2007;**434-435**:633–636

[72] Khor E, Lim LY. Implantable applications of chitin and chitosan. Biomaterials. 2003;**24**: 2339–2349

[73] Gupta AK, Wells S. Surface-modified superparamagnetic nanoparticles for drug delivery: Preparation, characterization, and cytotoxicity studies. IEEE Transactions on NanoBioscience. 2004;**3**:66–73

[74] Zhang Y, Kohler N, Zhang M. Surface modification of superparamagnetic magnetite nanoparticles and their intracellular uptake. Biomaterials. 2002;**23**:1553–1561

[75] Gupta AK, Curtis ASG. Surface modified superparamagnetic nanoparticles for drug delivery: Interaction studies with human fibroblasts in culture. Journal of Materials Science: Materials in Medicine. 2004;**15**:493–496

[76] Sudakar C, Dixit A, Regmi R, Naik R, Lawes G, Naik VM, Vaishnava PP, Toti U, Panyam J. Fe_3O_4 incorporated AOT-alginate nanoparticles for drug delivery. IEEE Transactions on Magnetics. 2008;**44**:2800–2803

[77] Ma HL, Xu YF, Qi XR, Maitani Y, Nagai T. Superparamagnetic iron oxide nanoparticles stabilized by alginate: Pharmacokinetics, tissue distribution, and applications in detecting liver cancers. International Journal of Pharmaceutics. 2008;**354**:217–226

[78] Morales MA, Finotelli PV, Coaquira JAH, Rocha-Leão MHM, Diaz-Aguila C, Baggio-Saitovitch EM, Rossi AM. In situ synthesis and magnetic studies of iron oxide nanoparticles in calcium-alginate matrix for biomedical applications. Materials Science Engineering: C. 2008;**28**:253–257

[79] Wang H, Luo W, Chen J. Fabrication and characterization of thermoresponsive Fe_3O_4@PNIPAM hybrid nanomaterials by surface-initiated RAFT polymerization. Journal of Materials Science. 2012;**47**:5918–5925

[80] Chen G, Hoffman AS. Preparation and properties of thermoreversible, phase-separating enzyme-oligo(N-isopropylacrylamide) conjugates. Bioconjugate Chemistry. 1993;**4**:509–514

[81] Regmi R, Bhattarai SR, Sudakar C, Wani AS, Cunningham R, Vaishnava PP, Naik R, Oupicky D, Lawes G. Hyperthermia controlled rapid drug release from thermosensitive magnetic microgels. Journal of Materials Chemistry. 2010;**20**:6158–6163

[82] Fang J, Wang C, Cao M, Cheng M, Shi J, Jin Y. Preparation and properties of multifunctional Fe_3O_4@PNIPAM-AAM@Au composites. Materials Letters. 2013;**96**:89–92

[83] Park I-K, Ng C-P, Wang J, Chu B, Yuan C, Zhang S, Pun SH. Determination of nanoparticle vehicle unpackaging by MR imaging of a T2 magnetic relaxation switch. Biomaterials. 2008;**29**:724–732

[84] Chorny M, Polyak B, Alferiev IS, Walsh K, Friedman G, Levy RJ. Magnetically driven plasmid DNA delivery with biodegradable polymeric nanoparticles. FASEB Journal. 2007;**21**:2510–2519

[85] Steitz B, Hofmann H, Kamau SW, Hassa PO, Hottiger MO, von Rechenberg B, Hofmann-Amtenbrink M, Petri-Fink A. Characterization of PEI-coated superparamagnetic iron oxide nanoparticles for transfection: Size distribution, colloidal properties and DNA interaction. Journal of Magnetism and Magnetic Materials. 2007;**311**:300–305

[86] McBain SC, Yiu HHP, El Haj A, Dobson J. Polyethyleneimine functionalized iron oxide nanoparticles as agents for DNA delivery and transfection. Journal of Materials Chemistry. 2007;**17**:2561–2565

[87] McBain SC, Yiu HHP, Dobson J. Magnetic nanoparticles for gene and drug delivery. International Journal of Nanomedicine. 2008;**3**:169–180

[88] Wang C, Ravi S, Martinez GV, Chinnasamy V, Raulji P, Howell M, Davis Y, Mallela J, Seehra MS, Mohapatra S. Dual-purpose magnetic micelles for MRI and gene delivery. Journal of Controlled Release. 2012;**163**:82–92

[89] Thevenot J, Oliveira H, Sandre O, Lecommandoux S. Magnetic responsive polymer composite materials. Chemical Society Review. 2013;**42**:7099–7116

[90] Karimi M, Ghasemi A, Sahandi Zangabad P, Rahighi R, Moosavi Basri SM, Mirshekari H, Amiri M, Shafaei Pishabad Z, Aslani A, Bozorgomid M, Ghosh D, Beyzavi A, Vaseghi A, Aref AR, Haghani L, Bahrami S, Hamblin MR. Smart micro/nanoparticles in stimulus-responsive drug/gene delivery systems. Chemical Society Review. 2016;**45**: 1457–1501

[91] Bonini M, Berti D, Baglioni P. Nanostructures for magnetically triggered release of drugs and biomolecules. Current Opinion in Colloid Interface Science. 2013;**18**:459–467

[92] Hayashi K, Ono K, Suzuki H, Sawada M, Moriya M, Sakamoto W, Yogo T. High-frequency, magnetic-field-responsive drug release from magnetic nanoparticle/organic hybrid based on hyperthermic effect. ACS Applied Materials Interfaces. 2010;**2**:1903–1911

[93] Munnier E, Cohen-Jonathan S, Linassier C, Douziech-Eyrolles L, Marchais H, Soucé M, Hervé K, Dubois P, Chourpa I. Novel method of doxorubicin-SPION reversible association for magnetic drug targeting. International Journal of Pharmaceutics. 2008;**363**:170–176

[94] Gautier J, Munnier E, Paillard A, Hervé K, Douziech-Eyrolles L, Soucé M, Dubois P, Chourpa I. A pharmaceutical study of doxorubicin-loaded PEGylated nanoparticles for magnetic drug targeting. International Journal of Pharmaceutics. 2012;**423**:16–25

[95] He X, Wu X, Cai X, Lin S, Xie M, Zhu X, Yan D. Functionalization of magnetic nanoparticles with dendritic-linear-brush-like triblock copolymers and their drug release properties. Langmuir. 2012;**28**:11929–11938

[96] Ying XY, Du YZ, Hong LH, Yuan H, Hu FQ. Magnetic lipid nanoparticles loading doxorubicin for intracellular delivery: Preparation and characteristics. Journal of Magnetism and Magnetic Materials. 2011;**323**:1088–1093

[97] Liao C, Sun Q, Liang B, Shen J, Shuai X. Targeting EGFR-overexpressing tumor cells using Cetuximab-immunomicelles loaded with doxorubicin and superparamagnetic iron oxide. European Journal of Radiology. 2011;**80**:699–705

[98] Quan Q, Xie J, Gao H, Yang M, Zhang F, Liu G, Lin X, Wang A, Eden HS, Lee S, Zhang G, Chen X. HSA coated iron oxide nanoparticles as drug delivery vehicles for cancer therapy. Molecular Pharmaceutics. 2011;**8**:1669–1676

[99] Gu YJ, Cheng J, Man CWY, Wong WT, Cheng SH. Gold-doxorubicin nanoconjugates for overcoming multidrug resistance. Nanomedicine: Nanotechnology. 2012;**8**:204–211

[100] Hua MY, Yang HW, Liu HL, Tsai RY, Pang ST, Chuang KL, Chang YS, Hwang TL, Chang YH, Chuang HC, Chuang CK. Superhigh-magnetization nanocarrier as a doxorubicin delivery platform for magnetic targeting therapy. Biomaterials. 2011;**32**:8999–9010

[101] Berry CC, Curtis AS. Functionalization of magnetic nanoparticles for applications in biomedicine. Journal of Physics D: Applied Physics. 2003;**36**:R198

[102] Chen F-H, Zhang L-M, Chen Q-T, Zhang Y, Zhang Z-J. Synthesis of a novel magnetic drug delivery system composed of doxorubicin-conjugated Fe_3O_4 nanoparticle cores

and a PEG-functionalized porous silica shell. Chemical Communications. 2010;**46**:8633–8635

[103] Nigam S, Chandra S, Newgreen DF, Bahadur D, Chen Q. Poly (ethylene glycol)-modified PAMAM-Fe3O4-doxorubicin triads with the potential for improved therapeutic efficacy: Generation-dependent increased drug loading and retention at neutral pH and increased release at acidic pH. Langmuir. 2014;**30**:1004–1011

[104] Akbarzadeh A, Samiei M, Joo SW, Anzaby M, Hanifehpour Y, Nasrabadi HT, Davaran S. Synthesis, characterization and in vitro studies of doxorubicin-loaded magnetic nanoparticles grafted to smart copolymers on A549 lung cancer cell line. Journal of Nanobiotechnology. 2012;**10**:46

[105] Arachchige MP, Laha SS, Naik AR, Lewis KT, Naik R, Jena BP. Functionalized nanoparticles enable tracking the rapid entry and release of doxorubicin in human pancreatic cancer cells. Micron. 2017;**92**:25–31

[106] Shokrollahi H. Contrast agents for MRI. Material Science Engineering: C. 2013;**33**:4485–4497

[107] Caravan P, Ellison JJ, McMurry TJ, Lauffer RB. Gadolinium (III) chelates as MRI contrast agents: Structure, dynamics, and applications. Chemical Reviews. 1999;**99**:2293–2352

[108] Arsalani N, Fattahi H, Nazarpoor M. Synthesis and characterization of PVP-functionalized superparamagnetic Fe_3O_4 nanoparticles as an MRI contrast agent. Express Polymer Letters. 2010;**4**:329–338

[109] Kim D, Zhang Y, Kehr J, Klason T, Bjelke B, Muhammed M. Characterization and MRI study of surfactant-coated superparamagnetic nanoparticles administered into the rat brain. Journal of Magnetism and Magnetic Materials. 2001;**225**:256–261

[110] Hu F, Wei L, Zhou Z, Ran Y, Li Z, Gao M. Preparation of biocompatible magnetite nanocrystals for in vivo magnetic resonance detection of cancer. Advance Materials. 2006;**18**:2553–2556

[111] Stephen ZR, Kievit FM, Zhang M. Magnetite nanoparticles for medical MR imaging. Material Today. 2011;**14**:330–338

[112] Comparative study of the magnetic behavior of spherical and cubic superparamagnetic iron oxide nanoparticles. Journal of Physics and Chemistry C. 2011;**115**:327–334

[113] Park YC, Smith JB, Pham T, Whitaker RD, Sucato CA, Hamilton JA, Bartolak-Suki E, Wong JY. Effect of PEG molecular weight on stability, T(2) contrast, cytotoxicity, and cellular uptake of superparamagnetic iron oxide nanoparticles (SPIONs). Colloids and Surfaces. B, Biointerfaces. 2014;**119**:106–114

[114] Xie J, Chen K, Lee H-Y, Xu C, Hsu AR, Peng S, Chen X, Sun S. Ultrasmall c(RGDyK)-coated Fe_3O_4 nanoparticles and their specific targeting to integrin $\alpha v\beta 3$-rich tumor cells. Journal of the American Chemical Society. 2008;**130**:7542–7543

[115] Huh Y-M, Jun Y-W, Song H-T, Kim S, Choi J-S, Lee J-H, Yoon S, Kim K-S, Shin J-S, Suh J-S, Cheon J. In vivo magnetic resonance detection of cancer by using multifunctional magnetic nanocrystals. Journal of the American Chemical Society. 2005;**127**:12387–12391

[116] Kim BH, Lee N, Kim H, An K, Park YI, Choi Y, Shin K, Lee Y, Kwon SG, Na HB, Park J-G, Ahn T-Y, Kim Y-W, Moon WK, Choi SH, Hyeon T. Large-scale synthesis of uniform and extremely small-sized iron oxide nanoparticles for high-resolution T1 magnetic resonance imaging contrast agents. Journal of the American Chemical Society. 2011;**133**: 12624–12631

[117] Jun Y-W, Huh Y-M, Choi J-S, Lee J-H, Song H-T, Kim Kim, Yoon S, Kim K-S, Shin J-S, Suh J-S, Cheon J. Nanoscale size effect of magnetic nanocrystals and their utilization for cancer diagnosis via magnetic resonance imaging. Journal of the American Chemical Society. 2005;**127**:5732–5733

[118] De M, Chou SS, Joshi HM, Dravid VP. Hybrid magnetic nanostructures (MNS) for magnetic resonance imaging applications. Advanced Drug Delivery Reviews. 2011;**63**: 1282–1299

[119] Shin T-H, Choi Y, Kim S, Cheon J. Recent advances in magnetic nanoparticle-based multi-modal imaging. Chemical Society Review. 2015;**44**:4501–4516

[120] Bae KH, Kim YB, Lee Y, Hwang J, Park H, Park TG. Bioinspired synthesis and characterization of gadolinium-labeled magnetite nanoparticles for dual contrast T1- and T2-weighted magnetic resonance imaging. Bioconjugate Chemistry. 2010;**21**:505–512

[121] Santra S, Jativa SD, Kaittanis C, Normand G, Grimm J, Perez JM. Gadolinium-encapsulating iron oxide nanoprobe as activatable NMR/MRI contrast agent. ACS Nano. 2012;**6**:7281–7294

[122] Zhou Z, Huang D, Bao J, Chen Q, Liu G, Chen Z, Chen X, Gao J. A synergistically enhanced T1–T2 dual-modal contrast agent. Advanced Materials. 2012;**24**:6223–6228

[123] Xiao N, Gu W, Wang H, Deng Y, Shi X, Ye L. T1–T2 dual-modal MRI of brain gliomas using PEGylated Gd-doped iron oxide nanoparticles. Journal of Colloid and Interface Science. 2014;**417**:159–165

[124] Maenosono S, Saita S. Theoretical assessment of FePt nanoparticles as heating elements for magnetic hyperthermia. IEEE Transactions on Magnetism. 2006;**42**:1638–1642

[125] Gilchrist RK, Medal R, Shorey WD, Hanselman RC, Parrott JC, Taylor CB. Selective inductive heating of lymph nodes. Annals of Surgery. 1957;**146**:596–606

[126] Sonvico F, Mornet S, Vasseur S, Dubernet C, Jaillard D, Degrouard J, Hoebeke J, Duguet E, Colombo P, Couvreur P. Folate-conjugated iron oxide nanoparticles for solid tumor targeting as potential specific magnetic hyperthermia mediators: Synthesis, physicochemical characterization, and in vitro experiments. Bioconjugate Chemistry. 2005;**16**: 1181–1188

[127] Sharifi I, Shokrollahi H, Amiri S. Ferrite-based magnetic nanofluids used in hyperthermia applications. Journal of Magnetism and Magnetic Materials. 2012;**324**:903–915

[128] Pereira C, Pereira AM, Fernandes C, Rocha M, Mendes R, Fernández-García MP, Guedes A, Tavares PB, Grenèche J-M, Araújo JOP. Superparamagnetic MFe_2O_4 (M= Fe, Co, Mn) nanoparticles: Tuning the particle size and magnetic properties through a novel one-step coprecipitation route. Chemistry of Materials. 2012;**24**:1496–1504

[129] Lasheras X, Insausti M, Gil de Muro I, Garaio E, Plazaola F, Moros M, De Matteis L, de la Fuente JSM, Lezama L. Chemical synthesis and magnetic properties of monodisperse nickel ferrite nanoparticles for biomedical applications. Journal of Physics and Chemistry C. 2016;**120**:3492–3500

[130] Dutz S, Kettering M, Hilger I, Müller R, Zeisberger M. Magnetic multicore nanoparticles for hyperthermia—Influence of particle immobilization in tumour tissue on magnetic properties. Nanotechnology. 2011;**22**:265102

[131] Vasseur S, Duguet E, Portier J, Goglio G, Mornet S, Hadová E, Knížek K, Maryško M, Veverka P, Pollert E. Lanthanum manganese perovskite nanoparticles as possible in vivo mediators for magnetic hyperthermia. Journal of Magnetism and Magnetic Materials. 2006;**302**:315–320

[132] Jang JT, Nah H, Lee JH, Moon SH, Kim MG, Cheon J. Critical enhancements of MRI contrast and hyperthermic effects by dopant-controlled magnetic nanoparticles. Angewandte Chemie. 2009;**121**:1260–1264

[133] Miller KJ, Sofman M, McNerny K, McHenry ME. Metastable γ-FeNi nanostructures with tunable Curie temperature. Journal of Applied Physics. 2010;**107**:09A305

[134] Nemala H, Thakur JS, Naik VM, Vaishnava PP, Lawes G, Naik R. Investigation of magnetic properties of Fe_3O_4 nanoparticles using temperature dependent magnetic hyperthermia in ferrofluids. Journal of Applied Physics. 2014;**116**:034309

[135] Rosensweig RE. Heating magnetic fluid with alternating magnetic field. Journal of Magnetism and Magnetic Materials. 2002;**252**:370–374

[136] Effect of nanoclustering and dipolar interactions in heat generation for magnetic hyperthermia. Langmuir. 2016;**32**:1201–1213

[137] Branquinho LC, Carrião MS, Costa AS, Zufelato N, Sousa MH, Miotto R, Ivkov R, Bakuzis AF. Effect of magnetic dipolar interactions on nanoparticle heating efficiency: Implications for cancer hyperthermia. Scientific Reports. 2013;**3**:2887

[138] Landi GT. Role of dipolar interaction in magnetic hyperthermia. Physics Review B. 2014;**89**:014403

[139] Sadat ME, Patel R, Sookoor J, Bud'ko SL, Ewing RC, Zhang J, Xu H, Wang Y, Pauletti GM, Mast DB, Shi D. Effect of spatial confinement on magnetic hyperthermia via dipolar interactions in Fe_3O_4 nanoparticles for biomedical applications. Materials Science Engineering: C. 2014;**42**:52–63

[140] Palihawadana-Arachchige M, Nemala H, Naik VM, Naik R. Effect of magnetic dipolar interactions on temperature dependent magnetic hyperthermia in ferrofluids. Journal of Applied Physics. 2017;**121**:023901

[141] Mohammad F, Balaji G, Weber A, Uppu RM, Kumar CS. Influence of gold nanoshell on hyperthermia of superparamagnetic iron oxide nanoparticles. Journal of Physics and Chemistry C. 2010;**114**:19194–19201

[142] Liu XL, Fan HM, Yi JB, Yang Y, Choo ESG, Xue JM, Fan DD, Ding J. Optimization of surface coating on Fe_3O_4 nanoparticles for high performance magnetic hyperthermia agents. Journal of Material Chemistry. 2012;**22**:8235–8244

[143] Cedervall T, Lynch I, Lindman S, Berggård T, Thulin E, Nilsson H, Dawson KA, Linse S. Understanding the nanoparticle–protein corona using methods to quantify exchange rates and affinities of proteins for nanoparticles. Proceedings of the National Academy of Sciences of the United States of America. 2007;**104**:2050–2055

[144] Klein J. Probing the interactions of proteins and nanoparticles. Proceedings of the National Academy of Sciences of the United States of America. 2007;**104**:2029–2030

[145] Khan S, Gupta A, Nandi CK. Controlling the fate of protein corona by tuning surface properties of nanoparticles. Journal of Physics and Chemistry Letters. 2013;**4**:3747–3752

[146] Calatayud MP, Sanz B, Raffa V, Riggio C, Ibarra MR, Goya GF. The effect of surface charge of functionalized Fe_3O_4 nanoparticles on protein adsorption and cell uptake. Biomaterials. 2014;**35**:6389–6399

[147] Khandhar AP, Ferguson RM, Krishnan KM. Monodispersed magnetite nanoparticles optimized for magnetic fluid hyperthermia: Implications in biological systems. Journal of Applied Physics. 2011;**109**:07B310

[148] Drug-loaded superparamagnetic iron oxide nanoparticles for combined cancer imaging and therapy in vivo. Angewendte Chemie International Edition. 2008;**47**:5362–5365

[149] Uniform mesoporous dye-doped silica nanoparticles decorated with multiple magnetite nanocrystals for simultaneous enhanced magnetic resonance imaging, fluorescence imaging, and drug delivery. Journal of the American Chemical Society. 2010;**132**:552–557.

[150] Kim J, Kim HS, Lee N, Kim T, Kim H, Yu T, Song IC, Moon WK, Hyeon T. Multifunctional uniform nanoparticles composed of a magnetite nanocrystal core and a mesoporous silica shell for magnetic resonance and fluorescence imaging and for drug delivery. Angewendte Chemie. 2008;**120**:8566–8569

[151] Thomas R, Park I-K, Jeong Y. Magnetic iron oxide nanoparticles for multimodal imaging and therapy of cancer. International Journal of Molecular Science. 2013;**14**:15910

[152] Hayashi K, Nakamura M, Sakamoto W, Yogo T, Miki H, Ozaki S, Abe M, Matsumoto T, Ishimura K. Superparamagnetic nanoparticle clusters for cancer theranostics combining magnetic resonance imaging and hyperthermia treatment. Theranostics. 2013;**3**:366–376

[153] Hilger I, Kaiser WA. Iron oxide-based nanostructures for MRI and magnetic hyperthermia. Nanomedicine. 2012;**7**:1443–1459

[154] Bozorth RM, Williams HJ, Walsh DE. Magnetic properties of some orthoferrites and cyanides at low temperatures. Physics Reviews. 1956;**103**:572–578

[155] Litsardakis G, Manolakis I, Efthimiadis K. Structural and magnetic properties of barium hexaferrites with Gd–Co substitution. Journal of Alloys and Compound. 2007;**427**:194–198

[156] Panda RN, Shih JC, Chin TS. Magnetic properties of nano-crystalline Gd- or Pr-substituted $CoFe_2O_4$ synthesized by the citrate precursor technique. Journal of Magnetism and Magnetic Materials. 2003;**257**:79–86

[157] Jiang P-S, Drake P, Cho H-J, Kao C-H, Lee K-F, Kuo C-H, Lin X-Z, Lin Y-J. Tailored nanoparticles for tumour therapy. Journal of Nanoscience and Nanotechnology. 2012;**12**:5076–5081

[158] Drake P, Cho H-J, Shih P-S, Kao C-H, Lee K-F, Kuo C-H, Lin X-Z, Lin Y-J. Gd-doped iron-oxide nanoparticles for tumour therapy via magnetic field hyperthermia. Journal of Materials Chemistry. 2007;**17**:4914–4918

[159] Peng J, Hojamberdiev M, Xu Y, Cao B, Wang J, Wu H. Hydrothermal synthesis and magnetic properties of gadolinium-doped $CoFe_2O_4$ nanoparticles. Journal of Magnetism and Magnetic Materials. 2011;**323**:133–137

[160] Palihawadana-Arachchige M, Naik V, Naik R. Gd doped Fe_3O_4 nanoparticles for magnetic hyperthermia and MRI. To be published.

Chapter 6

Dendrimer Structure Diversity and Tailorability as a Way to Fight Infectious Diseases

Dariusz T. Mlynarczyk, Tomasz Kocki and
Tomasz Goslinski

Additional information is available at the end of the chapter

http://dx.doi.org/10.5772/67660

Abstract

Dendrimers represent a distinct class of polymers—highly branched and uniform, with a relatively small size when compared to their mass. They are composed of the core, from which branched polymeric dendrons diverge and they are end-capped with selected terminal groups. Recently, dendrimers have attracted considerable attention from medicinal chemists, mostly due to their well-defined and easy-to-modify structure. This chapter aims to compile dendrimer applications and activities especially for prevention and fighting off infections caused by bacteria and fungi, viruses, and parasites/protozoa. Our goal in this review is to discuss selected modifications of dendrimers of potential value for pharmaceutical chemistry.

Keywords: antibacterial, antimycotic, dendrimer, nanomaterials, polymers

1. Introduction

Dendrimers are spherical nanosized polymers that branch in a well-defined manner. They were first synthesized and described by Franz Vögtle in 1978 [1]. For the last 30 years, they have gained a lot of attention, mainly due to the discovery of their stunning complexation abilities. In medicinal chemistry, they reveal interesting pharmacological properties of potential value for various medical fields. In pharmaceutical sciences, these nanostructures are particularly interesting as they can be potentially useful in pharmaceutical technology for preparation of water-soluble complexes with poorly soluble active pharmaceutical ingredients (API). It is worth noting that at the same time a decrease in API's overall toxicity is observed.

In this chapter, the aim is to describe the potential use of dendrimers in fighting off infectious diseases. Infectious diseases continue to constitute a problem around the globe, and a proper surveillance is required, as the amount of reports regarding occurrence of bacteria and fungi resistant to all clinically used antibiotics and antimycotics is growing. From one year to the next, the number of potentially useful antimicrobials is slowly decreasing, while only a few APIs have been introduced to clinical practice in recent years. Therefore, even treatment of common diseases has the potential to become a serious problem in the near future. The link between pollution and health although complex is obvious. An increasing pollution of the environment with pharmaceuticals intended to fight infectious diseases as well as their enhanced consumption over the last 70 years has led to the development of resistance mechanisms. Additionally, the treatment of infectious diseases in developing countries is quite problematic due to the lack of regulations in drug marketing. Another factor is the price of what is often referred to as last resort medicines; the contribution of public funding is often essential for the implementation of therapy with such medicinal products. Moreover, decreased hygiene level and underdeveloped sanitation favour the occurrence of bacterial, fungal and parasitic infections [2, 3].

Currently, the amount of novel antibiotic classes seems to be constant. In parallel, the propagation of multidrug resistant microbes indicates the necessity of searching for novel agents and methods, which can be used as a revolutionary approach. Therefore, present research wrestles with the problem of whether novel dendrimer nanoparticles alone or in complexes with APIs can be of potential usage in medicinal and pharmaceutical chemistry. The goal is to develop novel antibiotics, antimicrobial and antiparasitic/protozoa therapies. Currently, there are many ongoing attempts that aim at increasing the efficiency of strategies against bacteria and fungi in planktonic and biofilm modes of growth. There are researched methods that lead to biofilm growth inhibition, disruption or eradication. These approaches include APIs with new mechanisms of action, like enzymes, salts, metal nanoparticles, antibiotics, acids, plant extracts or antimicrobial photodynamic therapy. Regarding all this, dendrimers could be a material that might help to reach this goal [4].

2. Dendrimer structural versatility

Dendrimers do not form a uniform group based on their chemical structure. They are different from other dendritic structures such as dendronic and dendritic surfaces, dendronized polymers, dendriplexes and dendrigrafts. Schematic representation of dendrimeric nanoparticle, which constitutes the main subject of this short review, is presented in **Figure 1** [5]. Generally, dendrimers are composed of three elements: (*i*) a core, (*ii*) branched dendrons and (*iii*) terminal groups. They are most commonly synthesized using two different synthetic pathways: (*i*) divergent or (*ii*) convergent (**Figure 2**). In the divergent approach, firstly the synthesis of the functionalized core (or inner part of the dendrimer) takes part. Then this structure serves as a scaffold for the synthesis of higher generation dendrimers. Convergent approach relies on the synthesis of dendrons, which can be further attached to the earlier synthesized core. Some synthetic chemistry techniques, including *click chemistry* are especially useful for the synthesis of dendrimers [6].

http://dx.doi.org/10.5772/67660

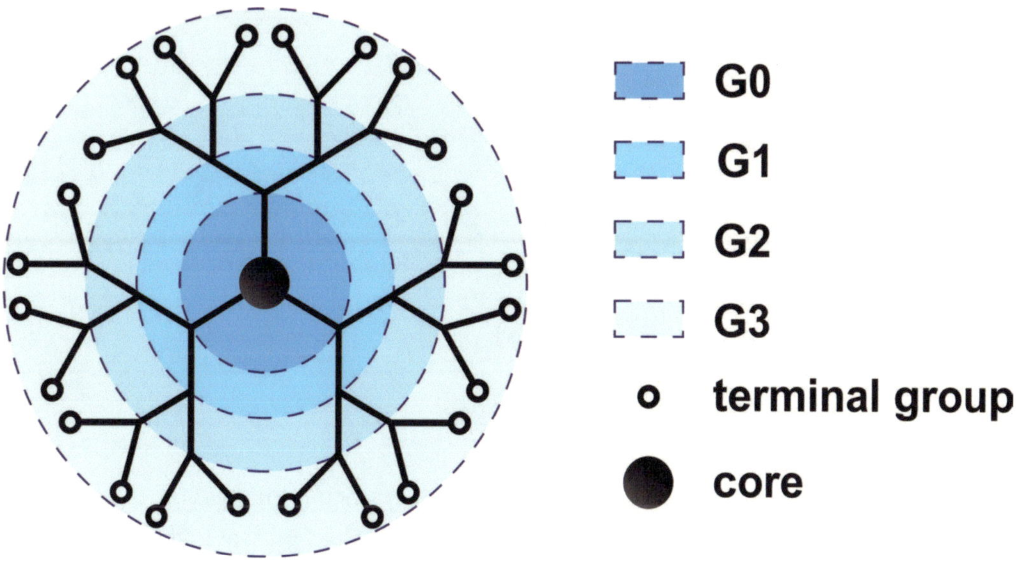

Figure 1. Schematic representation of dendrimer structure.

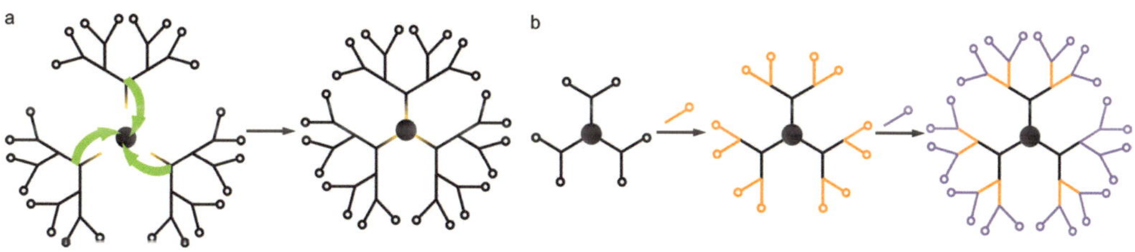

Figure 2. Schematic representation of convergent (a) and divergent (b) method of dendrimer synthesis.

A system was proposed describing the specific branching architecture of dendrimers, with general abbreviation AB_n—where n stands for new branches that arise from a node. For example, AB_2 and AB_3 states for two and three branches outgoing from each node, respectively. For graphical description, see **Figure 3** [6]. The most common dendrimers that can be found in the literature are built of AB_2 building blocks. Among those to the most popular belong **p**oly(**am**ido**am**ine)—PAMAM, poly(**p**ropylene**i**mine)—PPI (also called **po**ly(**p**ropylene**am**ine)—POPAM), carbosilane and **p**oly(**L-l**ysine)—PLL dendrimers [7].

The core part of dendrimers may be a variety of molecules—starting with single atoms (like nitrogen in some PAMAMs), aliphatic chains, alicyclic or aromatic rings through polyaromatic moieties, inorganic frameworks, and ending with other polymers and peptides. The core of a dendrimer can be simply a scaffold to which dendrons are attached. However, in some cases, the core is a molecule that expresses its own activity and added dendrons modify the periphery of the central molecule, thus affecting its physico-chemical properties (solubility, photochemical and electrochemical properties, protection from enzymes, etc.) [8–12].

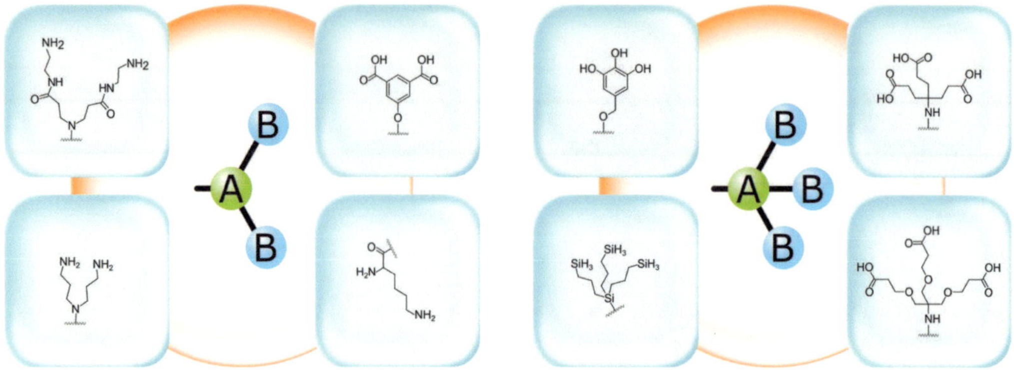

Figure 3. Schematic depiction of dendrimer branching modes with examples.

Dendrons serve mainly as carriers for other compounds. The controlled release of drugs from dendron-drug complexes can be modulated at certain pH values present in the environment of living organisms. An acidic environment is often associated with cancerous tissues, which was the subject of research by Wang *et al.* [13]. In this study, a potent and specific proteasome inhibitor bortezomib was released from biocompatible modified PAMAM dendrimer complex when triggered by the acidic environment of the tumour. End-group or terminal group of a dendrimer is a peripheral functionalization of dendrimer. The outstanding feature that makes dendrimers so special is the enormous amount of terminal groups in such a small molecule. The number of end-groups increases in exponential manner while the round molecule size increases linearly. This is depicted in **Table 1** for PAMAM dendrimer. Quite often the structure of end-groups depends on the synthetic method applied (harsh reaction conditions, protective groups) and the nature of the dendrimer building blocks [6].

Dendrimer end-groups can be easily modified. Modification changes their polarity and solubility in different solvents. In this way, high toxicity associated with many free amino groups in PAMAM and PPI dendrimers may be overcome by substituting them partially with non-toxic moieties [15, 16]. Alternatively, appending the end-groups with hydrophobic substituents may be considered, when they are intended to be utilized as carriers in hydrophobic formulations. In this way, the toxicity of prepared dendrimer is kept at bay and its complexation capabilities in hydrophobic mediums are increased. Utilizing this method, Hamilton *et al.* [17], while studying potential proteomimetics for ophthalmic use, decorated hydroxyl-terminated G5 PAMAM dendrimers with either dodecyl or cholesteryl moieties. Interestingly, use of dodecyl chains highly increased the overall molecule toxicity, as assessed on Chinese Hamster ovarian cells and pig lens epithelial cells. Cholesteryl-modified G5 PAMAM expressed toxicity to a lesser extent and was less toxic against lens epithelial cells than unmodified dendrimer. In another study, appending the hydrophobic Fréchet-type dendrimers with highly water-soluble carboxylic salt groups enabled the use of them in hydrophilic solvents [8].

Alternatively, the end-groups may be substituted with active substances, targeting molecules and others, that are relevant and needed for modern applications. Najlah and D'Emanuele reviewed the literature on the subject of dendrimer-drug conjugates [18]. The main benefit

http://dx.doi.org/10.5772/67660

Dendron generation	Number of end-groups			PAMAM dendrimer size [nm]	
	AB_2 dendron	AB_3 dendron	PAMAM dendrimer [7]	Calculated [14]	Experimental hydrodynamic diameter [7]
0	1	1	4	N/A	1.5
1	2	3	8	1.0	2.2
2	4	9	16	1.4	2.9
3	8	27	32	1.6	3.6
4	16	81	64	2.1	4.5
5	32	243	128	2.8	5.4
6	64	729	256	3.6	6.7
7	128	2187	512	4.5	8.1
8	256	6561	1024	5.7	9.7
9	512	19683	2048	6.8	11.4
10	1024	59049	4096	8.6	13.5

Table 1. Number of end-groups of dendrons for AB_2, AB_3 and PAMAM dendrimers and the size of PAMAM dendrimers.

from combining APIs with dendrimers in such manner is that dendrimer-API conjugates are more stable in various conditions as compared to their complexes based on non-covalent bonds. A good example for covalent bonding of APIs to dendrimer surface groups is the use of dendrimers as carriers for immunoactive peptides in the formation of vaccines. Such approach has been successfully tested by Skwarczynski *et al.* [19]. They have synthesized G1 triazole-based dendrons bond to a tetravalent core and with B-cell epitopes as end-groups. As this structure of amphiphillic properties was introduced to water solution, it assembled to form micelles with hydrophobic inner shell and hydrophilic—peptide—outer shell. Another type of terminal group modification concerns the use of targeting molecules, such as folic acid for cancer cells. Majoros *et al.* [20] designed and synthesized PAMAM dendrimer that was appended with several moieties—covalently bound drug (paclitaxel), target molecule (folic acid) and an imaging agent (fluorescein isothiocyanate). It is worth noting that not all terminal groups have to be modified at the same time. There are examples in the literature when the synthesis of dendrimers was performed in a statistical manner, so that only part of end-groups in dendrimer was modified, while the rest remained unchanged. Alternatively, convergent synthetic approach enables to combine various dendrons or polymers into one dendrimeric molecule. Such approach was successfully implemented by Albertazzi *et al.* [21]. Researchers managed to obtain hybrid dendritic-linear block copolymers based on a 4-arm and 2-arm polyethylene glycol (PEG) core. The 4-arm-based dendrimer showed significantly improved DNA binding and gene transfection capabilities in comparison with the 2-arm derivative.

The plethora of different modifications that can be proposed makes dendrimers perfect molecules for any chosen application with endless possibilities. For more insight into the potential applications of dendrimers and a broad spectrum of different properties of these nanosized polymers, the reader can refer to the comprehensive reviews such as Ref. [5] or Ref. [22]. Astruc *et al.* [5] broadly reviewed dendrimers designed for various functions and they discussed many examples of dendrimers and indicated how their physical, photophysical and supramolecular properties influence further applications in sensing, catalysis, molecular electronics, photonics and nanomedicine. The biomedical applications of dendrimers refer to the following areas: drug delivery, boron neutron capture therapy, photodynamic therapy, photothermal therapy based on gold and iron oxide nanoparticles, medical diagnostics and molecular probes (biosensors). It is worth noting that biomedical applications of dendrimers have gained a lot of attention during the last quarter of the century. In addition to dendrimers' own pharmacodynamic properties, especially promising is their possibility to be used as nanocarriers of improved solubility in water, biodistribution, ADME profile and pharmacokinetics with many other desired properties including extended circulation time and enhanced penetration and retention (EPR) property. Lately, dendrimers have been considered as agents of anti-amyloidogenic potential, as non-viral vectors of oligonucleotides and siRNA for gene therapy purposes and towards detection of IgE-mediated drug allergy reactions [5].

For pharmaceutical technology, dendrimers are mostly known for their carrier abilities. They exhibit great complexation potential for biologically active compounds, drugs, dyes and metal ions. Dendrimer carrier abilities for various chemical molecules (drugs, pigments, salts) comprise both drug encapsulation and chemical bonding to the periphery (**Figure 4**). Dendrimers have already found some commercial applications, for example, as a component of sexually transferred diseases preventing gel (*VivaGel*), a non-viral transfection agent (*Superfect*) and a contrast agent for MRI (*Gadomer 17*). Regardless of many potential uses and outstanding properties, there are also some drawbacks associated with dendrimers and their broader commercial use—these are usually time-consuming and expensive synthesis and purification of higher generations of dendrimers. Also high concentration of functional groups on the periphery debilitates its efficient modification [5, 18].

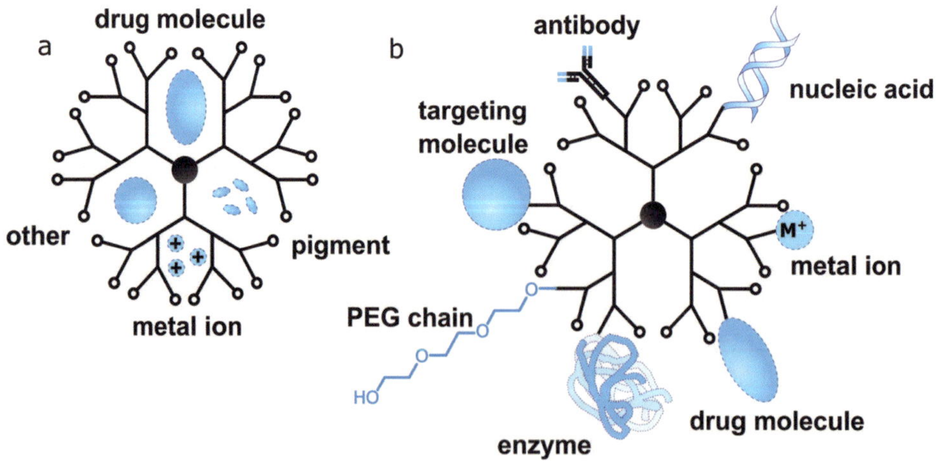

Figure 4. Dendrimer carrier abilities: (a) compound encapsulation and (b) covalent bonding of compound to the periphery.

http://dx.doi.org/10.5772/67660

3. Use of dendrimers against infectious diseases

3.1. Dendrimers as antibacterials and antifungals

Therapeutic efficiency of dendrimers as nanocarriers has been proved so far for, for example, potent anticancer, nonsteroidal and anti-inflammatory, antimicrobial and antiviral drugs. In this respect, two strategies have been applied for the application of dendrimers as drug carriers. The first one was encapsulation of drugs inside dendrimers or their binding to peripheral groups of dendrimers by electrostatic or ionic interactions. The second one concerned covalent bonding of drugs to the periphery of dendrimers [5]. The antibacterial activity of dendrimers has been already reviewed by Tülü and Ertürk [23] and Mintzer and co-workers [24]. The aim was to highlight the diversity of dendrimer structural modifications that led to an increased *per se* activity, as well as a decreased toxicity, thus prompting further applications in medical sciences.

Commercially available PAMAM dendrimers are effective antibacterial compounds (**Figure 5**). Amino-terminated G2 PAMAM dendrimer revealed differentiated minimum inhibitory concentration (MIC) and minimum bactericidal concentration (MBC) activities against various strains of *Escherichia coli*, *Staphylococcus aureus*, *Staphylococcus epidermidis*, *Salmonella enterica*, *Klebsiella pneumoniae* and MDR-*Shigella flexneri*. Moreover, it exhibited little toxicity to human gastric epithelial cells and did not induce antibiotic resistance in bacteria [25]. Unfortunately, due to many amino end-groups, G3 and higher generation PAMAMs were found to be highly toxic to living organisms [25, 26]. To counter this, PAMAM dendrimers were modified in other studies with hydroxyl end-groups or the syntheses were stopped at so-called half-generations, which resulted in ester group end-capped dendrons. It was noticed that such end-group modifications affect the solubility of the desired dendrimers, while greatly decreasing their toxicities and enabling their assessment for potential use in biological systems. Good examples in this regard are the studies on PAMAM end-group modifications by Calabretta and co-workers and Lopez and co-workers [15, 27]. In these studies, highly toxic G3 and

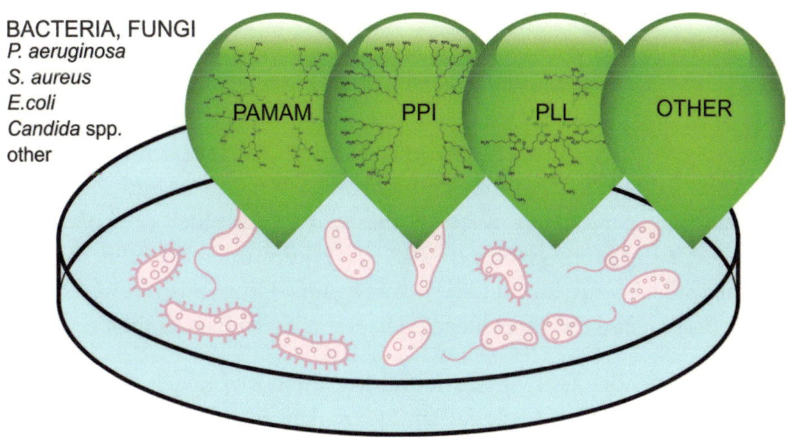

Figure 5. Schematic representation of dendrimer action on bacteria and fungi.

G5 PAMAM dendrimers with amino-terminated end-groups were partially substituted with poly(ethylene glycol) chains. Such PEGylated dendrimers expressed significantly lower toxicities, while antibacterial activities against *Pseudomonas aeruginosa* and *S. aureus* remained at a high level. In a similar manner, PAMAM dendrimers up to G4 were appended with aliphatic NO-releasing moieties [28, 29]. Such hybrids were tested on bacterial biofilms of *P. aeruginosa* [28, 29] and *S. aureus* [28] with positive results. It is worth noting that nitric oxide, which was additionally released, increased overall antibacterial activity. In addition, tested dendrimers revealed low-toxicity to mammalian cells.

A highly active but also toxic dendrimer is G4 PPI. Felczak *et al.* co-workers overcame its toxicity by attaching maltose to 25 or 100% of all the amino end-groups [16]. The modified maltose-appended PPI dendrimers were microbiologically assessed against *S. aureus, S. epidermidis, E. coli, P. aeruginosa, Candida albicans* and found to be active against both bacteria and fungi with the lowered toxicity assessed against a Chinese hamster fibroblast cell line (B14), human liver hepatocellular carcinoma cell line (HepG2), mouse neuroblastoma cell line (N2a) and rat liver cell line (BRL-3A). The PPI G4 dendrimer modified with 25% of maltose was found superior to the 100% modified one in terms of antibacterial activity and demonstrated a striking selectivity towards *S. aureus* at the concentrations non-toxic for eukaryotic cell lines. As the continuation of this study, the G4 PPI dendrimer appended in 25% with maltose was used in combination with nadifloxacin against Gram-negative bacteria: *E. coli, P. aeruginosa, Proteus hauseri* revealing an increase in the antibacterial activity of the latter. Similar to the previous studies, the tested fluoroquinolone in dendrimer-complex was found to be less toxic than the drug alone [30]. In another study, unmodified G4 PPI complex with ciprofloxacin was found to increase quinolone activity at non-toxic concentrations to mammalian cells [31]. It is worth pointing out that other quinolones (nadifloxacin and prulifloxacin) were also complexed with dendrimers, like PAMAM (up to the 5th generation) [32]. Although, this combination did not enhance the potency of these drugs against *E. coli*, an increase in their solubility was observed.

Another type of dendrimers, poly(phosphorhydrazone) dendrimers appended with PEG chains were synthesized on a solid support provided by silica nanoparticles [33]. These composites were used for hosting silver-based nanoparticles and assessed on the basis of their antibacterial activity, which was found to reach Gram-negative (*E. coli, P. aeruginosa*) and Gram-positive (*S. aureus, Enterococcus hirae*) bacteria. Wang *et al.* conducted a similar experiment, although utilizing titanium dioxide-supported G5 PAMAMs that were also PEGylated [34]. However, no silver addition was needed as the titania-supported composite tested as the thin antibacterial film was found to inhibit growth of *S. aureus* and *P. aeruginosa.* Again, phosphorus-containing dendrimers were synthesized by Ciepluch *et al.* [35] who obtained viologen-phosphorus dendrons (viologen—4,4′-bipyridinium salts) built around a cyclotriphosphazene core, which were found to be active against Gram positive (*S. aureus*) and Gram negative (*E. coli, P. aeruginosa, Proteus vulgaris*) bacteria, as well as against *C. albicans.*

An amino acid-based dendrimer [36] was found to exhibit low toxicity and high antibacterial activity against usually resistant bacterial strains of *Acinetobacter baumanii* and *P. aeruginosa,* including multidrug-resistant and extensively drug-resistant strains. Abd-El-Aziz *et al.* published the synthesis of G0-G2 dendrimers built of melamine core, arene-cyclopentadienyliron-based dendrons and piperazine end-groups [37]. These dendrimers were assessed against drug-resistant bacteria methicillin-resistant *S. aureus* (MRSA), *Staphylococcus warneri* and vancomycin-resistant *Enterococcus faecium* (VRE). It is noted that the G2 piperazine end-capped dendrimer was the most active. Interesting in its structure was polyanionic ether-based G1 dendrimer with lipophilic core, which was highly active against Gram-positive bacteria—*Bacillus subtillis* [38]. The noteworthy point is that this dendrimer was found to exhibit low toxicity to healthy human cells. Carbosilane dendrimers have also been found to be potent antimicrobials. Fuentes-Paniagua *et al.* synthesized and investigated microbiological evaluation of dendrimers up to the 2nd generation with different terminal groups, which revealed promising MIC and MBC values against *S. aureus* and *E. coli* [39].

Quite a lot of attention has also been given to so-called antimicrobial peptides. These are short, naturally occurring peptides that exhibit high antimicrobial activity. Problems associated with the use of these compounds are related to their susceptibility to bacterial enzymes and a not fully recognized mechanism of action. Reports published on this subject regarding antibacterial activity of synthetic short dendrimeric peptides suggest the high potential of such an approach. Lind *et al.* synthesized modified amino acid-based low-generation dendrimers [40]. The compounds obtained in this study were assessed for their action against *S. aureus, E. coli* and *P. aeruginosa* and proven active. At the same time, some hemotoxicity was observed. Structurally similar peptide appended PLL dendrimers were the subject of a study performed by Bruschi *et al.* [41] who synthesized a functionalized dendrimer that showed promising potential as an antibacterial agent interacting with bacterial lipopolysaccharide. It exhibited quite good activity against various Gram-positive strains and especially high activity against Gram-negative strains with potency similar to lipopeptide antibiotics like colistin and polymixin B. One of the most promising activities of dendrimers against pathogenic fungi was reported by Staniszewska *et al.* [42] who designed and synthesized an amino acid-based dendrimer end-capped with tyrosine moieties for specific action against fungi. This dendrimer was found not only to be active against reference and clinical strains but also affected the virulence factors of *Candida* species.

Another approach of utilizing dendrimers in the fight against bacteria is to combine them with other structures or compounds. In this way, PAMAMs were combined with multiwalled carbon nanotubes and CdS or Ag_2S quantum dots to form novel hybrid materials [43]. Nanohybrids were found to be highly active against bacteria and the activity was found to be just as high or higher than for each component alone. As a continuation of this study, authors functionalized the surface of multiwalled carbon nanotubes with polyamide dendrons [44]. This composite material was used for synthesis of silver nanoparticles and then applied as a carrier for these. Organic–inorganic hybrid was assessed and proven effective as antimicrobial against *E. coli, P. aeruginosa* and *S. aureus*. PAMAM dendrimers were also numerously tested with metals and their ions. The G2 and G3 PAMAMs were utilized as a site for synthesis of silver nanoparticles [45]. After grafting cotton with G2 PAMAM-nanosilver complex, the material exhibited bactericidal activity against *S. aureus* and *E. coli*. In other studies, utilizing polyester dendrimers for nano-silver formation, it was shown that the antibacterial activity of the resulting dendrimer-silver nanoparticles was strongly dependent on the dendrimer generation [46]. Noteworthy is that nanoparticles that were stabilized by dendrimers exhibited no toxic effect on human epithelial cell line A549 as has been proven using 3-(4,5-dimethylthiazol-2-yl)-2,5-diphenyltetrazolium bromide (MTT) assay. Another nanosized composite was developed by Strydom *et al.* [47]. They synthesized silver sulfadiazine nanoparticles stabilized with the addition of G1, G3.5, G4 or G4.5 PAMAM dendrimers. Such nanoparticles in topical formulations were evaluated in terms of their antibacterial activity. The use of dendrimers improved the activity of resulting material against *Staphylococcus* spp., *E. coli* and *P. aeruginosa*. Another interesting work was published by Staneva and co-workers [48]. They managed to append G1 PPI dendrimer with naphthalimide derivative. This nanostructure was found to create complex with bivalent ions, such as zinc and copper. Both metal ion complexes exhibited activity against Gram-negative and Gram-positive bacteria, as well as pathogenic fungi. A further application of dendrimers was proposed by Zainul Abid and co-workers, who applied PAMAM for water disinfection [49]. The antibacterial activity (*E. coli, S. aureus, P. aeruginosa, Bacillus subtilis*) of the G1 dendrimer was increased by polymerizing it with ethyleneglycol dimethacrylate to form polydendrimeric network. The resultant polymer exhibited very high antibacterial activity at 10 mg/L in a matter of minutes. The amount of 100% bacteria was eradicated within 2 minutes of incubation. Gangadharan *et al.* utilized dendrimers for water disinfection synthesizing them on a solid support [50]. They applied a divergent synthetic pathway to dendrimeric poly-ethylene amine supported on polystyrene copolymer beads and assessed its potential activity against both Gram-positive (*S. aureus, B. subtilis*) and negative (*E. coli, P. aeruginosa*) bacteria. A decrease in bacterial cell viability was seen, yet the effect was strongly dependent on the generation of supported dendrimers and their end-group functionalization.

Almost all dendrimers described in this chapter derive their high antibacterial and antifungal activity from the so-called starburst effect. Dendrimers mentioned earlier are characterized by the exponential growth of the number of terminal groups. Such a rapid increase in the number of active sites of small molecules (by means of their volume) is the result of this phenomenon. In case of dendrimers as antimicrobial drug carriers, dendrimeric formulations are often just as effective or even more so on pathogens as a drug used alone. Use of dendrimers usually results in prolonged release of the drug with simultaneously decreased toxicity comparing to the parent compound. This can be clearly seen for a plethora of drug molecules. For example, such a study was performed for sulfamethoxazole, which is a poorly soluble sulfonamide. Its solubility increased in the formulations prepared with PAMAM dendrimers [51]. This increase was generation-dependent. As the result of this change, an increase of sulfamethoxazole antibacterial activity and sustained release of the drug were observed. Navath *et al.* carried out a study with amoxicillin in modified G4 PAMAM formulation, which was found to be non-toxic and enabled a sustained release of the drug *in vitro* [52]. The study was enhanced to *in vivo* assessment on pregnant guinea pigs. The intravaginal dendrimer-amoxicillin biodegradable formulation was found not to cause such adverse effects as necrosis or inflammatory response in deeper tissues. Also, no pH change in application site was observed nor was any transfer across the foetal membranes noticed. Amoxicillin was also the subject of the report by Wrońska *et al.* [53]. However, in this case the G3 PPI, either unmodified or maltose end-capped was studied. Because of toxicity related to unmodified PPI dendrimers, results obtained for maltose terminated PPIs are of significance. Antibacterial activity of drugs increased by the application of the above mentioned combination. Another drug molecule of antimicrobial activity that was successfully solubilized by PAMAMs was triclosan [54]. Although this formulation has not been tested on bacteria, the amount of drug solubilized was remarkable and—as found in this report—highly dependent on the pH of the analysed solution. Various interesting topical formulations were prepared by Wroblewska and Winnicka with the use of erythromycin and G2 or G3 PAMAM dendrimers [55]. The formulations were optimized and their utility for therapy proven by *in vitro* assessment against *Enterococcus faecalis* and different strains of *S. aureus*. The same group continued their research with erythromycin and PAMAMs by attempting to increase the antibiotic water-solubility [56]. In this study, PAMAM dendrimers were used to solubilize erythromycin and tobramycin to form stable water solutions. While this goal was achieved, a slight increase in antibacterial activity against Gram-positive bacteria (in this case, *S. aureus* and *E. faecalis*) by means of MBC of solubilized erythromycin was observed. However, tobramycin used with dendrimers exhibited unchanged or lower MICs and MBCs when compared to the drug alone. One of the latest reports describes the development of the G4 PAMAM containing lipid hybrid nanoparticle capable of transporting one of the most potent last-resort antibiotics—vancomycin [57]. The results clearly indicate that the use of dendrimer carrier increases vancomycin potency against MRSA. Moreover, Skwarczynski *et al.* developed a G1 triazole-based dendrimers, which were end-capped with B-cell epitope of *Streptococcus pyogenes* [19]. Such potential vaccine for *Straphyloccus* infection was found to be self-adjuvating by forming micelle-like assemblies with dendrimer core, branches in the centre and peptides on the periphery. Apart from of immunization properties, the authors suggest, that such approach may be applied to other peptides in the preparation of vaccines for various diseases.

3.2. Dendrimers for treatment and prevention of virus-related infection

Dendrimers have been applied for treatment and prevention of virus-related infections (**Figure 6**). The best-known dendrimer acting in an antiviral manner is probably SPL7013, discovered by Starpharma, which is the active ingredient of VivaGel [58]. It is a G4 PLL dendrimer with functionalized end-groups used in the form of a gel and marketed as condom lubrication. SPL7013 successfully underwent second-stage clinical trials and was found to prevent sexually transmitted diseases, most notably *Human immunodeficiency virus 1* (HIV-1), *Herpes simplex virus* (HSV) and *Human papilloma virus* (HPV). Recently, Starpharma claimed to obtain results indicating prevention of Zika virus infection. Sánchez-Rodríguez *et al.* synthesized second-generation carbosilane dendrimers, functionalized on the periphery with highly anionic sulfonate groups or carboxy groups [59]. Both synthesized dendrimers when applied to the vagina-derived human epithelial VK2/E6E7 cell lines and peripheral blood mononuclear cells before exposure to HIV decreased its infectivity. The test conditions also included physiological pH to prove their potential utility in protection against *in vivo* sexually transmitted HIV infection. Authors discussed the superior activity of carbosilane dendrimers as compared to PAMAMs, as carbosilane dendrimers do not induce inflammatory reaction, which influences the infectivity of HIV. Structurally related compounds were also the subject of investigation by Vacas-Córdoba and co-workers [60]. They synthesized the G1 carbosilane dendrimer appended with sulfonated naphthyl groups and G2 carbosilane dendrimer with sulphated end-groups and then assessed their action on HIV and its infectivity. The results suggested that developed dendrimers decrease the infectivity rate of pre-treated peripheral

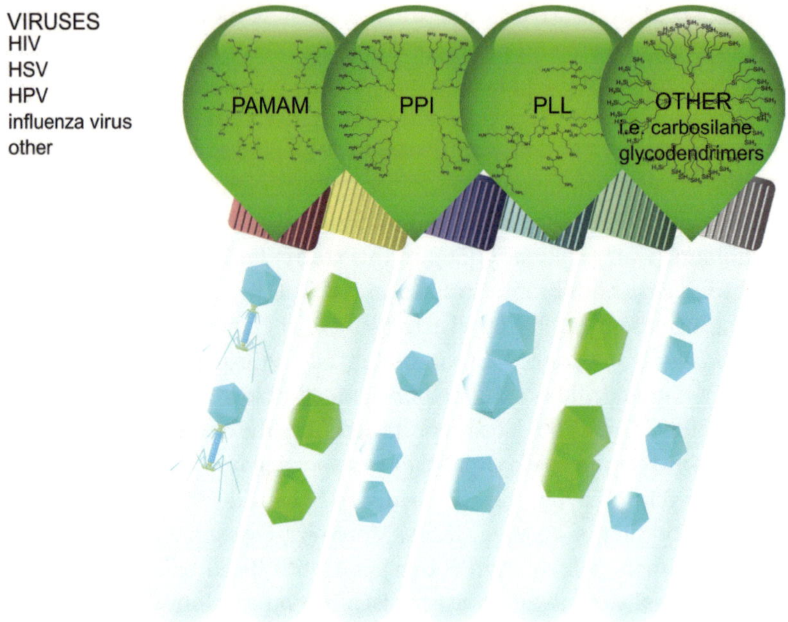

Figure 6. Schematic representation of dendrimer action on viruses.

blood mononuclear cells when exposed to the virus. Authors concluded that poly-anionic dendrimers act by electrostatic interaction with HIV proteins inhibiting the ability of the virus to bind to target cells and thus its ability to internalize. In addition, the cell-to-cell transmission was impaired, as assessed in a different test by co-culturing infected cells with healthy TZM. bl cells (modified HeLa cell line especially prone to HIV infection). High capability to prevent infection by the G1 dendrimer was also presented in the presence of seminal fluid [61].

Carbosilane dendrimers were also investigated in terms of other potential antiviral applications [62, 63]. Knowing that dendrimers are excellent non-viral transfection agents, two polycationic G1 carbosilane dendrimers with different cores were assessed for gene therapy in order to inhibit the development of ongoing HIV infection. Both dendrimers were found to exhibit properties making them suitable for their planned use. Their non-toxicity was confirmed (MTT assay). They were able to form complexes with nucleic acids and—as siRNA complexes—inhibit replication of HIV-1 and affect macrophage response, thus encouraging further study on this subject.

Antiviral PPI dendrimers were also assessed for potential anti-HIV treatment [64]. In this case, dendrimers up to their third generation were modified on the periphery with anionic groups such as carboxylate or sulfonate functional groups. Modified dendrimers were used to complex bivalent metal ions: Cu^{2+}, Ni^{2+}, Co^{2+} or Zn^{2+}. Metal complexes were assessed for their HIV infection potential applying *in vitro* models of first and second infection barriers. Complexes exhibited high inhibition rates and prevented internalization of HIV-1. Inhibition of the virus replication was also observed. Han *et al.* synthesized a G3 PLL built on the core of ammonia and functionalized with sulphated cellobiose as terminal groups [65]. The dendrimer was found to be non-toxic and active against HIV, as revealed by *in vitro* MTT assay. Replacing ammonia core with long alkyl stearylamide chain resulted in obtaining dendrimer with comparable EC_{50} values [66]. Recently Ceña-Díez *et al.* published a review paper dealing with the topic of various dendrimers expressing antiviral activity against heparan sulphate-related viral diseases caused by HSV, HIV, HPV, *Hepatitis C virus* and *Human respiratory syncytial virus* [67]. The authors concluded that all the dendrimers analysed (i.e. amino acid-based dendrimers, glycodendrimers, PAMAMs, carbosilane dendrimers) were acting by means of preventing the virus entry to the cell.

Dendrimers have been also assessed as potential vaccine carriers. For an excellent review on this subject, the paper by Heegaard *et al.* is recommended [68]. A very interesting study on this subject was performed by Chahal *et al.* who modified a G1 PAMAM dendrimer with long aliphatic chains [69]. Such functionalized dendrimer among other vaccine components was used to prepare a formulation for transporting mRNA chains. In this way, a vaccine was developed that was capable of providing protection with a one-dose application from H1N1 influenza virus, *Toxoplasma gondii* and Ebola virus. This was confirmed by performing *in vivo* experiment and by comparing the results with bare mRNA vaccination. PAMAM formulation was proven to increase stability of the vaccine by protecting nucleic acids from RNAses. The vaccine was stable even after 30 days of storage. In another study, a vaccine against rabies was developed by Ullas *et al.* [70] who used G3 poly-ether imine dendrimer to prepare a formulation of rabies virus glycoprotein gene. The efficiency of the formulation was evaluated by measuring the rabies virus neutralizing antibody levels in treated mice. Addition of dendrimer resulted in 4.5-fold greater increase and thus all the mice that underwent rabies virus challenge survived, while in the group treated with formulation without dendrimer only 60% viability was observed. PLL dendrimers were the subject of research by Blanco *et al.* to create a suitable vaccination for swine from foot-and-mouth disease virus [71]. They prepared G0 and G1 PLL dendrimers with T-epitope as a core and B-epitopes as terminal groups. Such prepared functionalized dendrimer was found to offer 100% protection in virus challenge.

A quite different approach was undertaken by Yandrapu *et al.* [72] who exploited the drug delivering properties of dendrimers. The G3.5 PAMAM was modified in the periphery with cysteamine. Modified dendrimers were loaded with acyclovir, a known antiviral, and assessed *in vitro* by means of drug release and mucoadhesive properties of the formulation. As in most cases, use of dendrimer formulation resulted in sustained release of the drug. Although unmodified dendrimer was characterized by higher acyclovir loading, presence of thiol groups was crucial for increase of bio-adhesion. In a different study, G4 PPI dendrimers were modified in the periphery by mannosylation and conjugation with sialic acid [73]. Functionalized dendrimer was used as a carrier for zidovudine. This complex was found to be more effective than PPI functionalized with only sialic acid or mannose and was also found non-toxic as well as zidovudine being released in a prolonged manner. Dendrimers with antimicrobial activity and for vaccination are also of commercial value as can be seen by a number of patents on the subject [74].

3.3. Dendrimers in fighting off the parasitic/protozoa infection

The action of dendrimers on protozoa and parasitic infections is mostly unexplored areas in comparison with their antiviral and antibacterial activity (**Figure 7**). There is also only limited data on the drug delivery of compounds for treating such infections using dendrimer carriers. Below is a summary and discussion of some reports on this subject.

Heredero-Bermejo *et al.* published a report regarding anti-trophozoite *Acanthamoeba castellanii* activity of G0 and G1 carbosilane dendrimers with ammonium salts as terminal groups [75]. It was found that G1 dendrimer had the best effect on *Acanthamoeba*, with LD_{50} at 0.193 μg/μL after 24 hours of exposure.

Sulfadiazine is one of the drugs used for treatment of *T. gondii* infections. For a more efficient method of this drug delivery, Prieto *et al.* assessed the possibility of using PAMAMs for its delivery [76]. They used commercially available G4 and G4.5 PAMAM dendrimers and loaded them with sulfadiazine. Such complexes were assessed by means of their drug loading capabilities, toxicity and *in vitro* activity. Each dendrimer was complexing up to 30 molecules of drug. Dendriplexes decreased the infection, with the highest decrease for G4 PAMAM-sulfadiazine at 30 nM concentration. In toxicity assessment, the G4.5 complex was found non-toxic, whilst G4 (amino end-capped) was proven toxic at low concentrations, as shown in MTT assay with Vero cells and murine macrophage-like cell line J774. In other studies, researchers focused on Amphotericin B (AmB), which is a drug that possesses anti-leishmanial activity. Jain *et al.* [77] evaluated the G5 PPI with muramyl dipeptide-modified terminal groups for its potential use as AmB carrier for treatment of leishmaniosis. Such formulation was compared to AmB commercial formulations: *Fungizone* and *AmBisome* assessing its toxicity and antiparasitic activity. The results indicated that this dendrimer significantly reduced the haemolytic toxicity of AmB to human erythrocytes, as well as cytotoxicity against J774A.1 macrophage cell line as shown

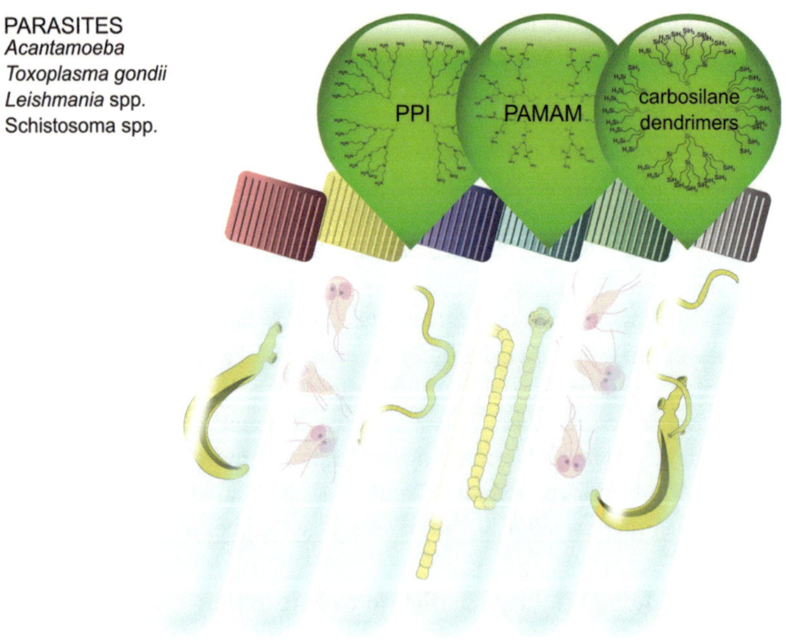

Figure 7. Schematic representation of dendrimer action on parasitic/protozoa infections.

by results of MTT assay. Simultaneously, G5 PAMAM-AmB formulation outruns commercial formulations, as can be seen in tests against *Leishmania donovani*. Jain *et al.* continued their efforts for finding a dendrimer capable of delivering AmB [78]. They reported G5 PPI dendrimer partially end-capped with mannose groups. Dendrimer revealed good loading capacity and AmB release was pH-dependent, as faster release was observed in acidic conditions. Haemolytic toxicity (human erythrocytes) and cytotoxicity (J774A.1 macrophage cell lines in MTT assay) of sole dendrimer and AmB-loaded dendrimer were assessed and the obtained data revealed a decreased toxicity of AmB-dendriplexes, negligible haemotoxicity and no cytotoxicity of dendrimer. Experiments on BALB/c mice were then conducted to determine *in vivo* properties: formulation pharmacokinetics, biodistibution, haematological toxicity and nephrotoxicity. These studies demonstrated that G5 PPI-mannose conjugate is a low toxic AmB formulation, providing prolonged release of carried drug. AmB was also the object of interest for Daftarian *et al.* [79]. They prepared G5 PAMAM bound to a targeting peptide—Pan-DR-binding epitope. Subsequently, such modified dendrimer was complexed with liposomal AmB in different ratios. Formation of a desired complex was confirmed by DLS analysis and electron microscopy. Cytotoxicity was tested on Hep2 cells and it was proven to be low. An *in vitro* and later *in vivo* study on infected BALB/c and C57BL mice was performed to assess therapeutic efficacy against *Leishmania major* infection. In the anti-leishmanial activity assessment, a six-fold decrease in AmB dose for the same therapeutic effect was observed. Due to the designed peptide terminal groups, AmB was more likely to accumulate in targeted cells.

It is worth noting that Wang *et al.* prepared a vaccine for *Schistosoma japonicum* infection [80]. To achieve this, they have end-capped a G4 PAMAM dendrimer with lysine, which greatly decreased the cytotoxicity to human kidney transformed cells 293T. The G4 PAMAM-lysine was then used as a carrier for DNA vaccine, namely SjC23, and *in vivo* assessment on BALB/c mice was conducted as a *S. japonicum* challenge after three prior immunizations. Dendrimer-based vaccine was found to reduce worm levels up to 50% and liver eggs by up to 62%.

3.4. Dendrimers for sensing of infective microbes

Dendrimers were also considered as components of microbial sensing devices (**Figure 8**). Use thereof as bacterial and viral presence has been reviewed by Satija *et al.* [81]. There is still not much known regarding fungi sensing, as only few reports have been published to date. Castillo *et al.* developed a system of indirect detection of fungal presence by sensing aflatoxin B_1—a known fungal toxin [82]. The system was built of gold electrode coupled with sequential monolayers of cystamine, glutaraldehyde, G4-PAMAM, glutaraldehyde and aptamers (synthetic receptors). Dendrimers used to construct this sensor provided the signal amplification, enabling the detection of aflatoxin B_1 in the concentration range 0.1–10 nmol. Use of different mycotoxins proved that the detection system was specific only for aflatoxin B_1 as the response was concentration-dependent in this case only. The second and last report regarding fungi sensing with the help of dendrimers was published by Mejri-Omrani *et al.* [83]. The authors described a system prepared for detection of ochratoxin A. The sensor was prepared with the use of G4 PAMAM dendrimers and found to detect ochratoxin A in concentration up to 5 μg/L with detection limit 2 ng/L. Developed sensor utility was confirmed in real food products tests.

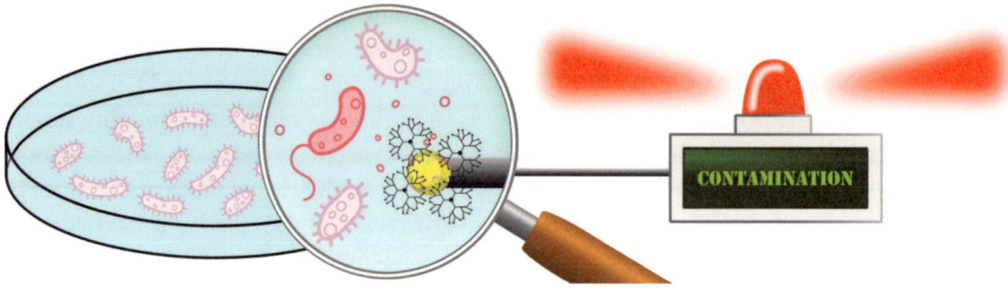

Figure 8. Schematic representation of dendrimer sensing.

Use of dendrimers for sensing parasitic presence was assessed by Perinoto *et al.* who developed a device suitable for sensing *Leishmania* infections in infected BALB/c mice serum samples [84]. The sensor was built by depositing in sequence several layers of G4 PAMAMs and liposomes containing antigenic proteins on a gold electrode. The device was able to detect *Leishmania amazonensis* antibodies in concentration as low as 10^{-5} mg/mL and the specificity of the sensor allowed it to distinguish between leishmaniasis and *Trypanosoma cruzi* infection.

4. Conclusions and perspectives

In recent years, dendrimers, which represent a distinct class of polymers, have gained considerable attention from medicinal chemists and pharmaceutical technologists, mostly due to their known and potential applications for medicine. Although there are still many unresolved issues on the topic of dendrimers, our goal in this review was to discuss selected modifications of dendrimers dedicated to the prevention and fighting of infections caused by bacteria, fungi, viruses and parasites.

A great increase of dendrimer-related research is to some extent bound to their commercial availability (mostly PAMAMs) as well as novel and efficient methods of their synthesis. In this regard, the development and commercial availability of various innovative building blocks for the synthesis of full-grown dendrimers is especially important. Dendrimer chemistry continues to develop year by year and many research groups and companies are interested in studying their properties and potential uses. In this chapter, many potential and practical applications of dendrimers in prevention of diseases, diagnostics of microbes have been discussed.

Based on the reviewed literature, dendrimers have proven to be useful in many ways. The starburst effect of nanopolymers obtained is magnified exponentially, resulting in unprecedented outcomes. A plethora of various studies on PAMAMs revealed how the dendrimers activity and toxicity changes upon a slight modification in their structure. Formation of dendriplexes quite often increases the biological activity of both dendrimer and encapsulated drug molecules. Furthermore, discoveries in the field of dendrimers encourage various studies in pharmaceutical technology on poor water-soluble APIs, which otherwise would not be considered useful for clinical practice. Moreover, there is still much to be done in regard to Gram-negative bacteria. Because of differences in cell wall structure, they are not as susceptible to anti-bacterials and disinfectants as their Gram-positive counterparts. In addition, studies published on dendrimers do not point out their mechanism of action. Based on the non-specific action of dendrimers on bacteria and fungi, one cannot assume that dendrimers always exhibit an identical mechanism of action on these organisms.

Only a few reports deal with the subject of infective fungi and there is almost none regarding the parasitic and protozoan infections. Systemic fungal infections are on the rise in developed countries because of the increasing use of immunosuppressive drugs after transplantations and due to opportunistic infections associated with HIV infected people suffering from AIDS. In terms of viruses, the research on the application of dendrimers for the improvement of vaccines is very important. As for parasites, encouraging are studies aimed at developing diagnostic methods for the detection of these organisms. Generally, high possibilities of modifying dendrimer core, branches and terminal groups, as well as development of methods on combining them with other active moieties make them unique and highly promising molecules for future use.

Acknowledgements

Authors thank the National Science Centre, Poland for funding (grant no. 2012/05/E/NZ7/01204). Authors would like to thank Mrs Agata Kaluzna-Mlynarczyk for her help with the graphics.

Author details

Dariusz T. Mlynarczyk[1], Tomasz Kocki[2] and Tomasz Goslinski[1*]

*Address all correspondence to: tomasz.goslinski@ump.edu.pl

1 Department of Chemical Technology of Drugs, Poznan University of Medical Sciences, Poznan, Poland

2 Department of Experimental and Clinical Pharmacology, Medical University of Lublin, Lublin, Poland

References

[1] Buhleier E, Wehner W, Vögtle F. "Cascade"- and "nonskid-chain-like" syntheses of molecular cavity topologies. Synthesis 1978; 1978:155-8. doi:10.1055/s-1978-24702

[2] Braine T. Race against time to develop new antibiotics. Bull World Health Organ 2011; 89: 88-9. doi:10.2471/BLT.11.030211

[3] Hof H. Will resistance in fungi emerge on a scale similar to that seen in bacteria? Eur J Clin Microbiol Infect Dis 2008; 27: 327-34. doi:10.1007/s10096-007-0451-9

[4] Collins TL, Markus EA, Hassett DJ, Robinson JB. The effect of a cationic porphyrin on *Pseudomonas aeruginosa* biofilms. Curr Microbiol 2010; 61: 411-6. doi:10.1007/s00284-010-9629-y

[5] Astruc D, Boisselier E, Ornelas C. Dendrimers designed for functions: from physical, photophysical, and supramolecular properties to applications in sensing, catalysis, molecular electronics, photonics, and nanomedicine. Chem Rev 2010; 110: 1857-959. doi:10.1021/cr900327d

[6] Campagna S, editor. Designing dendrimers. Hoboken, NJ: Wiley; 2012.

[7] Fréchet JMJ, Tomalia DA, editors. Dendrimers and other dendritic polymers. John Wiley & Sons, Ltd; Baffins Lane, Chichester, West Sussex, 2001.

[8] Setaro F, Ruiz-González R, Nonell S, Hahn U, Torres T. Synthesis, photophysical studies and $^{1}O_2$ generation of carboxylate-terminated zinc phthalocyanine dendrimers. J Inorg Biochem 2014; 136: 170-6. doi:10.1016/j.jinorgbio.2014.02.007

[9] Wieczorek E, Piskorz J, Popenda L, Jurga S, Mielcarek J, Goslinski T. First example of a diazepinoporphyrazine with dendrimeric substituents. Tetrahedron Lett 2017; 58: 758-761. doi:10.1016/j.tetlet.2017.01.027

[10] Mlynarczyk DT, Lijewski S, Falkowski M, Piskorz J, Szczolko W, Sobotta L, et al. Dendrimeric sulfanyl porphyrazines: synthesis, physico-chemical characterization, and biological activity for potential applications in photodynamic therapy. ChemPlusChem 2016; 81: 460-70. doi:10.1002/cplu.201600051

[11] Falkowski M, Rebis T, Piskorz J, Popenda L, Jurga S, Mielcarek J, et al. Improved electrocatalytic response toward hydrogen peroxide reduction of sulfanyl porphyrazine/multiwalled carbon nanotube hybrids deposited on glassy carbon electrodes. Dyes Pigm 2016; 134: 569-79. doi:10.1016/j.dyepig.2016.08.014

[12] Tillo A, Stolarska M, Kryjewski M, Popenda L, Sobotta L, Jurga S, et al. Phthalocyanines with bulky substituents at non-peripheral positions: synthesis and physico-chemical properties. Dyes Pigm 2016; 127: 110-5. doi:10.1016/j.dyepig.2015.12.017

[13] Wang M, Wang Y, Hu K, Shao N, Cheng Y. Tumor extracellular acidity activated "off–on" release of bortezomib from a biocompatible dendrimer. Biomater Sci 2015; 3: 480-9. doi:10.1039/C4BM00365A

[14] Maiti PK, Çagın T, Wang G, Goddard WA. Structure of PAMAM dendrimers: generations 1 through 11. Macromolecules 2004; 37: 6236-54. doi:10.1021/ma035629b

[15] Calabretta MK, Kumar A, McDermott AM, Cai C. Antibacterial activities of poly(amidoamine) dendrimers terminated with amino and poly(ethylene glycol) groups. Biomacromolecules 2007; 8: 1807-11. doi:10.1021/bm0701088

[16] Felczak A, Wrońska N, Janaszewska A, Klajnert B, Bryszewska M, Appelhans D, et al. Antimicrobial activity of poly(propylene imine) dendrimers. New J Chem 2012; 36: 2215. doi:10.1039/c2nj40421d

[17] Hamilton PD, Jacobs DZ, Rapp B, Ravi N. Surface hydrophobic modification of fifth-generation hydroxyl-terminated poly(amidoamine) dendrimers and its effect on biocompatibility and rheology. Materials 2009; 2: 883-902. doi:10.3390/ma2030883

[18] Najlah M, D'Emanuele A. Synthesis of dendrimers and drug-dendrimer conjugates for drug delivery. Curr Opin Drug Discov Devel 2007; 10: 756-67.

[19] Skwarczynski M, Zaman M, Urbani CN, Lin I-C, Jia Z, Batzloff MR, et al. Polyacrylate dendrimer nanoparticles: a self-adjuvanting vaccine delivery system. Angew Chem Int Ed 2010; 49: 5742-5. doi:10.1002/anie.201002221

[20] Majoros IJ, Myc A, Thomas T, Mehta CB, Baker JR. PAMAM dendrimer-based multifunctional conjugate for cancer therapy: synthesis, characterization, and functionality. Biomacromolecules 2006; 7: 572-9. doi:10.1021/bm0506142

[21] Albertazzi L, Mickler FM, Pavan GM, Salomone F, Bardi G, Panniello M, et al. Enhanced bioactivity of internally functionalized cationic dendrimers with PEG cores. Biomacromolecules 2012; 13: 4089-97. doi:10.1021/bm301384y

[22] Klajnert B, Peng L, Cena V, editors. Dendrimers in biomedical applications; RSC Publishing, Cambridge, 2013.

[23] Tülü M, Ertürk AS. Dendrimers as antibacterial agents 2012. In: Bobbarala V, editor. A search for antibacterial agents. 1st ed. Rijeka: Intech. p. 89-106. doi:10.5772/46051

[24] Mintzer MA, Dane EL, O'Toole GA, Grinstaff MW. Exploiting dendrimer multivalency to combat emerging and re-emerging infectious diseases. Mol Pharm 2012; 9: 342-54. doi:10.1021/mp2005033

[25] Xue X, Chen X, Mao X, Hou Z, Zhou Y, Bai H, et al. Amino-terminated generation 2 poly(amidoamine) dendrimer as a potential broad-spectrum, nonresistance-inducing antibacterial agent. AAPS J 2013; 15: 132-42. doi:10.1208/s12248-012-9416-8

[26] Pryor JB, Harper BJ, Harper SL. Comparative toxicological assessment of PAMAM and thiophosphoryl dendrimers using embryonic zebrafish. Int J Nanomed 2014; 9: 1947-56. doi:10.2147/IJN.S60220

[27] Lopez AI, Reins RY, McDermott AM, Trautner BW, Cai C. Antibacterial activity and cytotoxicity of PEGylated poly(amidoamine) dendrimers. Mol Biosyst 2009; 5: 1148. doi:10.1039/b904746h

[28] Worley BV, Schilly KM, Schoenfisch MH. Anti-biofilm efficacy of dual-action nitric oxide-releasing alkyl chain modified poly(amidoamine) dendrimers. Mol Pharm 2015; 12: 1573-83. doi:10.1021/acs.molpharmaceut.5b00006

[29] Lu Y, Slomberg DL, Shah A, Schoenfisch MH. Nitric oxide-releasing amphiphilic poly(amidoamine) (PAMAM) dendrimers as antibacterial agents. Biomacromolecules 2013; 14: 3589-98. doi:10.1021/bm400961r

[30] Felczak A, Zawadzka K, Wrońska N, Janaszewska A, Klajnert B, Bryszewska M, et al. Enhancement of antimicrobial activity by co-administration of poly(propylene imine) dendrimers and nadifloxacin. New J Chem 2013; 37: 4156. doi:10.1039/c3nj00760j

[31] Wrońska N, Felczak A, Zawadzka K, Janaszewska A, Klajnert B, Bryszewska M, et al. The antibacterial effect of the co-administration of poly(propylene imine) dendrimers and ciprofloxacin. New J Chem 2014; 38: 2987. doi:10.1039/c3nj01338c

[32] Cheng Y, Qu H, Ma M, Xu Z, Xu P, Fang Y, et al. Polyamidoamine (PAMAM) dendrimers as biocompatible carriers of quinolone antimicrobials: an in vitro study. Eur J Med Chem 2007; 42: 1032-8. doi:10.1016/j.ejmech.2006.12.035

[33] Hameau A, Collière V, Grimoud J, Fau P, Roques C, Caminade A-M, et al. PPH dendrimers grafted on silica nanoparticles: surface chemistry, characterization, silver colloids hosting and antibacterial activity. RSC Adv 2013; 3: 19015. doi:10.1039/c3ra43348j

[34] Wang L, Erasquin UJ, Zhao M, Ren L, Zhang MY, Cheng GJ, et al. Stability, antimicrobial activity, and cytotoxicity of poly(amidoamine) dendrimers on titanium substrates. ACS Appl Mater Interfaces 2011; 3: 2885-94. doi:10.1021/am2004398

[35] Ciepluch K, Katir N, Kadib A El, Felczak A, Zawadzka K, Weber M, et al. Biological properties of new viologen-phosphorus dendrimers. Mol Pharm 2012; 9: 448-57. doi:10.1021/mp200549c

[36] Pires J, Siriwardena TN, Stach M, Tinguely R, Kasraian S, Luzzaro F, et al. *In vitro* activity of the novel antimicrobial peptide dendrimer G3KL against multidrug-resistant *Acinetobacter baumannii* and *Pseudomonas aeruginosa*. Antimicrob Agents Chemother 2015; 59: 7915-8. doi:10.1128/AAC.01853-15

[37] Abd-El-Aziz AS, Abdelghani AA, El-Sadany SK, Overy DP, Kerr RG. Antimicrobial and anticancer activities of organoiron melamine dendrimers capped with piperazine moieties. Eur Polym J 2016; 82: 307-23. doi:10.1016/j.eurpolymj.2016.04.002

[38] Meyers SR, Juhn FS, Griset AP, Luman NR, Grinstaff MW. Anionic amphiphilic dendrimers as antibacterial agents. J Am Chem Soc 2008; 130: 14444-5. doi:10.1021/ja806912a

[39] Fuentes-Paniagua E, Hernández-Ros JM, Sánchez-Milla M, Camero MA, Maly M, Pérez-Serrano J, et al. Carbosilane cationic dendrimers synthesized by thiol-ene click chemistry and their use as antibacterial agents. RSC Adv 2014; 4: 1256-65. doi:10.1039/C3RA45408H

[40] Lind T, Polcyn P, Zielinska P, Cárdenas M, Urbanczyk-Lipkowska Z. On the antimicrobial activity of various peptide-based dendrimers of similar architecture. Molecules 2015; 20: 738-53. doi:10.3390/molecules20010738

[41] Bruschi M, Pirri G, Giuliani A, Nicoletto SF, Baster I, Scorciapino MA, et al. Synthesis, characterization, antimicrobial activity and LPS-interaction properties of SB041, a novel dendrimeric peptide with antimicrobial properties. Peptides 2010; 31: 1459-67. doi:10.1016/j.peptides.2010.04.022

[42] Staniszewska M, Bondaryk M, Zielińska P, Urbańczyk-Lipkowska Z. The in vitro effects of new D186 dendrimer on virulence factors of *Candida albicans*. J Antibiot (Tokyo) 2014; 67: 425-32.

[43] Neelgund GM, Oki A, Luo Z. Antimicrobial activity of CdS and Ag_2S quantum dots immobilized on poly(amidoamine) grafted carbon nanotubes. Colloids Surf B Biointerfaces 2012; 100: 215-21. doi:10.1016/j.colsurfb.2012.05.012

[44] Neelgund GM, Oki A. Deposition of silver nanoparticles on dendrimer functionalized multiwalled carbon nanotubes: synthesis, characterization and antimicrobial activity. J Nanosci Nanotechnol 2011; 11: 3621-9. doi:10.1166/jnn.2011.3756

[45] Tang J, Chen W, Su W, Li W, Deng J. Dendrimer-encapsulated silver nanoparticles and antibacterial activity on cotton fabric. J Nanosci Nanotechnol 2013; 13: 2128-35. doi:10.1166/jnn.2013.6883

[46] Mahltig B, Cheval N, Astachov V, Malkoch M, Montanez MI, Haase H, et al. Hydroxyl functional polyester dendrimers as stabilizing agent for preparation of colloidal silver particles—a study in respect to antimicrobial properties and toxicity against human cells. Colloid Polym Sci 2012; 290: 1413-21. doi:10.1007/s00396-012-2650-x

[47] Strydom SJ, Rose WE, Otto DP, Liebenberg W, de Villiers MM. Poly(amidoamine) dendrimer-mediated synthesis and stabilization of silver sulfonamide nanoparticles with increased antibacterial activity. Nanomed Nanotechnol Biol Med 2013; 9: 85-93. doi:10.1016/j.nano.2012.03.006

[48] Staneva D, Vasileva-Tonkova E, Makki MSI, Sobahi TR, Abdel-Rahman RM, Boyaci IH, et al. Synthesis and spectral characterization of a new PPA dendrimer modified with 4-bromo-1,8-naphthalimide and in vitro antimicrobial activity of its Cu(II) and Zn(II) metal complexes. Tetrahedron 2015; 71: 1080-7. doi:10.1016/j.tet.2014.12.083

[49] Zainul Abid CKV, Jackeray R, Jain S, Chattopadhyay S, Asif S, Singh H. Antimicrobial efficacy of synthesized quaternary ammonium polyamidoamine dendrimers and dendritic polymer network. J Nanosci Nanotechnol 2016; 16: 998-1007. doi:10.1166/jnn.2016.10656

[50] Gangadharan D, Dhandhala N, Dixit D, Thakur RS, Popat KM, Anand PS. Investigation of solid supported dendrimers for water disinfection. J Appl Polym Sci 2012; 124: 1384-91. doi:10.1002/app.34967

[51] Ma M, Cheng Y, Xu Z, Xu P, Qu H, Fang Y, et al. Evaluation of polyamidoamine (PAMAM) dendrimers as drug carriers of anti-bacterial drugs using sulfamethoxazole (SMZ) as a model drug. Eur J Med Chem 2007; 42: 93-8. doi:10.1016/j.ejmech.2006.07.015

[52] Navath RS, Menjoge AR, Dai H, Romero R, Kannan S, Kannan RM. Injectable PAMAM dendrimer–PEG hydrogels for the treatment of genital infections: formulation and *in vitro* and *in vivo* evaluation. Mol Pharm 2011; 8: 1209-23. doi:10.1021/mp200027z

[53] Wrońska N, Felczak A, Zawadzka K, Poszepczyńska M, Różalska S, Bryszewska M, et al. Poly(propylene imine) dendrimers and amoxicillin as dual-action antibacterial agents. Molecules 2015; 20: 19330-42. doi:10.3390/molecules201019330

[54] Gardiner J, Freeman S, Leach M, Green A, Alcock J, D'Emanuele A. PAMAM dendrimers for the delivery of the antibacterial triclosan. J Enzyme Inhib Med Chem 2008; 23: 623-8. doi:10.1080/14756360802205257

[55] Wróblewska M, Winnicka K. The effect of cationic polyamidoamine dendrimers on physicochemical characteristics of hydrogels with erythromycin. Int J Mol Sci 2015; 16: 20277-89. doi:10.3390/ijms160920277

[56] Winnicka K, Wroblewska M, Wieczorek P, Sacha P, Tryniszewska E. The effect of PAMAM dendrimers on the antibacterial activity of antibiotics with different water solubility. Molecules 2013; 18: 8607-17. doi:10.3390/molecules18078607

[57] Sonawane SJ, Kalhapure RS, Rambharose S, Mocktar C, Vepuri SB, Soliman M, et al. Ultra-small lipid-dendrimer hybrid nanoparticles as a promising strategy for antibiotic delivery: in vitro and in silico studies. Int J Pharm 2016; 504: 1-10. doi:10.1016/j.ijpharm.2016.03.021

[58] ©Starpharma Holdings Limited. Accessed January 8, 2017 from http://www.starpharma.com/vivagel

[59] Sánchez-Rodríguez J, Díaz L, Galán M, Maly M, Gómez R, la Mata FJ de, et al. Anti-human immunodeficiency virus activity of thiol-ene carbosilane dendrimers and their potential development as a topical microbicide. J Biomed Nanotechnol 2015; 11: 1783-98. doi:10.1166/jbn.2015.2109

[60] Vacas-Córdoba E, Maly M, la Mata FJ de, Gómez R, Pion M, Munoz-Fernandez MA. Antiviral mechanism of polyanionic carbosilane dendrimers against HIV-1. Int J Nanomed 2016;11: 1281-1294. doi:10.2147/IJN.S96352

[61] Ceña Diez R, García Broncano P, la Mata FJ de, Gómez R, Munoz-Fernandez MA. Efficacy of HIV antiviral polyanionic carbosilane dendrimer G2-S16 in the presence of semen. Int J Nanomed 2016;11: 2443-2450. doi:10.2147/IJN.S104292

[62] Weber N, Ortega P, Clemente MI, Shcharbin D, Bryszewska M, la Mata FJ de, et al. Characterization of carbosilane dendrimers as effective carriers of siRNA to HIV-infected lymphocytes. J Control Release 2008; 132: 55-64. doi:10.1016/j.jconrel.2008.07.035

[63] Perisé-Barrios AJ, Jiménez JL, Domínguez-Soto A, la Mata FJ de, Corbí AL, Gomez R, et al. Carbosilane dendrimers as gene delivery agents for the treatment of HIV infection. J Control Release 2014; 184: 51-7. doi:10.1016/j.jconrel.2014.03.048

[64] García-Gallego S, Díaz L, Jiménez JL, Gómez R, la Mata FJ de, Muñoz-Fernández MÁ. HIV-1 antiviral behavior of anionic PPI metallo-dendrimers with EDA core. Eur J Med Chem 2015; 98: 139-48. doi:10.1016/j.ejmech.2015.05.026

[65] Han S, Yoshida D, Kanamoto T, Nakashima H, Uryu T, Yoshida T. Sulfated oligosaccharide cluster with polylysine core scaffold as a new anti-HIV dendrimer. Carbohydr Polym 2010; 80: 1111-5. doi:10.1016/j.carbpol.2010.01.031

[66] Han S, Kanamoto T, Nakashima H, Yoshida T. Synthesis of a new amphiphilic glycodendrimer with antiviral functionality. Carbohydr Polym 2012; 90: 1061-8. doi:10.1016/j.carbpol.2012.06.044

[67] Ceña-Díez R, Sepúlveda-Crespo D, Maly M, Muñoz-Fernández MA. Dendrimeric based microbicides against sexual transmitted infections associated to heparan sulfate. RSC Adv 2016; 6: 46755-64. doi:10.1039/C6RA06969J

[68] Heegaard PMH, Boas U, Sorensen NS. Dendrimers for vaccine and immunostimulatory uses. A review. Bioconjug Chem 2010; 21: 405-18. doi:10.1021/bc900290d

[69] Chahal JS, Khan OF, Cooper CL, McPartlan JS, Tsosie JK, Tilley LD, et al. Dendrimer-RNA nanoparticles generate protective immunity against lethal Ebola, H_1N_1 influenza, and *Toxoplasma gondii* challenges with a single dose. Proc Natl Acad Sci 2016; 113: E4133-42.

[70] Ullas PT, Madhusudana SN, Desai A, Sagar BKC, Jayamurugan G, Rajesh YRD, et al. Enhancement of immunogenicity and efficacy of a plasmid DNA rabies vaccine by nanoformulation with a fourth-generation amine-terminated poly(ether imine) dendrimer. Int J Nanomed 2014; 9: 627. doi:10.2147/IJN.S53415

[71] Blanco E, Guerra B, la Torre BG de, Defaus S, Dekker A, Andreu D, et al. Full protection of swine against foot-and-mouth disease by a bivalent B-cell epitope dendrimer peptide. Antiviral Res 2016; 129: 74-80. doi:10.1016/j.antiviral.2016.03.005

[72] Yandrapu SK, Kanujia P, Chalasani KB, Mangamoori L, Kolapalli RV, Chauhan A. Development and optimization of thiolated dendrimer as a viable mucoadhesive excipient for the controlled drug delivery: an acyclovir model formulation. Nanomed Nanotechnol Biol Med 2013; 9: 514-22. doi:10.1016/j.nano.2012.10.005

[73] Gajbhiye V, Ganesh N, Barve J, Jain NK. Synthesis, characterization and targeting potential of zidovudine loaded sialic acid conjugated-mannosylated poly(propyleneimine) dendrimers. Eur J Pharm Sci 2013; 48: 668-79. doi:10.1016/j.ejps.2012.12.027

[74] Heegaard PM, Boas U. Dendrimer based anti-infective and anti-inflammatory drugs. Recent Patents Anti-Infect Drug Disc 2006; 1: 333-51.

[75] Heredero-Bermejo I, Copa-Patiño JL, Soliveri J, García-Gallego S, Rasines B, Gómez R, et al. In vitro evaluation of the effectiveness of new water-stable cationic carbosilane dendrimers against *Acanthamoeba castellanii* UAH-T17c3 trophozoites. Parasitol Res 2013; 112: 961-9. doi:10.1007/s00436-012-3216-z

[76] Prieto MJ, Bacigalupe D, Pardini O, Amalvy JI, Venturini C, Morilla MJ, et al. Nanomolar cationic dendrimeric sulfadiazine as potential antitoxoplasmic agent. Int J Pharm 2006; 326: 160-8. doi:10.1016/j.ijpharm.2006.05.068

[77] Jain K, Verma AK, Mishra PR, Jain NK. Characterization and evaluation of amphotericin B loaded MDP conjugated poly(propylene imine) dendrimers. Nanomed Nanotechnol Biol Med 2015; 11: 705-13. doi:10.1016/j.nano.2014.11.008

[78] Jain K, Verma AK, Mishra PR, Jain NK. Surface-engineered dendrimeric nanoconjugates for macrophage-targeted delivery of amphotericin B: formulation development and *in vitro* and *in vivo* evaluation. Antimicrob Agents Chemother 2015; 59: 2479-87. doi:10.1128/AAC.04213-14

[79] Daftarian PM, Stone GW, Kovalski L, Kumar M, Vosoughi A, Urbieta M, et al. A targeted and adjuvanted nanocarrier lowers the effective dose of liposomal amphotericin B and enhances adaptive immunity in murine cutaneous leishmaniasis. J Infect Dis 2013; 208: 1914-22. doi:10.1093/infdis/jit378

[80] Wang X, Dai Y, Zhao S, Tang J, Li H, Xing Y, et al. PAMAM-Lys, a novel vaccine delivery vector, enhances the protective effects of the SjC23 DNA vaccine against *Schistosoma japonicum* infection. PLoS One 2014; 9: e86578. doi:10.1371/journal.pone.0086578

[81] Satija J, Sai VVR, Mukherji S. Dendrimers in biosensors: concept and applications. J Mater Chem 2011; 21: 14367. doi:10.1039/c1jm10527b

[82] Castillo G, Spinella K, Poturnayová A, Šnejdárková M, Mosiello L, Hianik T. Detection of aflatoxin B1 by aptamer-based biosensor using PAMAM dendrimers as immobilization platform. Food Control 2015; 52: 9-18. doi:10.1016/j.foodcont.2014.12.008

[83] Mejri-Omrani N, Miodek A, Zribi B, Marrakchi M, Hamdi M, Marty J-L, et al. Direct detection of OTA by impedimetric aptasensor based on modified polypyrrole-dendrimers. Anal Chim Acta 2016; 920: 37-46. doi:10.1016/j.aca.2016.03.038

[84] Perinoto ÂC, Maki RM, Colhone MC, Santos FR, Migliaccio V, Daghastanli KR, et al. Biosensors for efficient diagnosis of leishmaniasis: innovations in bioanalytics for a neglected disease. Anal Chem 2010; 82: 9763-8. doi:10.1021/ac101920t

Chapter 7

Polymer Nanocomposites

Maria Inês Bruno Tavares, Emerson Oliveira da Silva,
Paulo Rangel Cruz da Silva and
Lívia Rodrigues de Menezes

Additional information is available at the end of the chapter

http://dx.doi.org/10.5772/intechopen.68142

Abstract

The development of polymer nanocomposites has advanced, especially due to their new properties after nanoparticle incorporation. Many nanocomposites composed of synthetic polymers and/or biopolymers have been studied after incorporation of a diversity of nanoparticles, which differ in form, shape, surface area and chemical organization. In this chapter, some examples of nanocomposites based on poly-vinyl alcohol (PVA); polycarbonate (PC) and matrixes of dental resins are presented. These nanocomposites could be obtained by three basic methods: in situ polymerization, solution casting and melt extrusion. The best method is determined by the relation and route to the polymer-nanoparticle pair. The dispersion and distribution of nanoparticles in the polymer matrix is the key to obtaining new materials with synergism of compounds properties. This synergism depends on how strong is the intermolecular interaction between the polymer matrix and nanoparticles. The evaluation of new nano systems can be done by different techniques, usually microscopy, X-ray diffraction, thermal analysis and so on. Low-field NMR relaxometry has been used to evaluate polymer nanocomposites. This technique provides valuable information related to the interaction of the nanoparticles with the polymer matrix, and it also indicates the dispersion and distribution of these nanoparticles in the matrix.

Keywords: nanocomposites, polymer, spin lattice relaxation

1. Introduction

Development of polymer nanocomposites has advanced quickly because these new materials normally present better properties when compared to pure polymers and/or polymer composites, due to the new architecture. Many natural polymers, synthetic polymers, biopolymers, and elastomers have been used to prepare these materials, containing different nanoparticles incorporated in them, depending on the application [1–9]. To obtain these new materials with good characteristics, it is necessary to choose the right polymer-nanoparticle pair and preparation technique, because the new architecture depends on this.

To generate polymer nanocomposites, the following important information on the components need to be considered in order to understand the behavior of the new materials: (i) polymer mass; (ii) polymer chemical structure; (iii) polymer semi-crystallinity; (iv) polymer chemical solubility; (v) polymer thermal stability; (vi) nanoparticle surface area; (vii) nanoparticle chemical structure; and (viii) nanoparticle dispersion. There are several methods to obtain these materials, the most common being *in situ* polymerization, solution dispersion (including nanoprecipitation and spray drying) and melt extrusion. Each process has its own particularity. But the essence of all polymer nanocomposites is the final morphology, irrespective of the process, which depends on polymer-nanoparticle interactions that will promote good dispersion and distribution of the nanoparticles in the polymer matrix [10–14]. The final morphology also depends on the process of obtaining the polymer nanocomposite.

2. Different methods for synthesis of polymer nanocomposites

2.1. *In situ* polymerization

This method normally is suitable for polymers that cannot be produced economically or safely by solution methods because the solvents used to dissolve them are highly toxic. This method promotes good dispersion and distribution of the nanoparticles in the polymer matrix [10, 15]. Some important aspects of this method should be pointed out. The first is related to the cost of the process, which can require some changes compared to the normal polymer synthesis. Care is also necessary to choose the most appropriate catalyst. The apparatus used can be the same for polymerization without nanoparticles.

2.2. Solution method

This is a good method when the solvent used is less toxic (chloroform, acetone, alcohol or water). In this method, different quantities of nanoparticles can be dispersed due to the good interaction with the solvent and polymer. This is the easiest method to obtain good nanocomposites. Some care must be taken in the manipulation of the solvent since it must be completely eliminated afterward [8, 11, 13, 16]. The necessary apparatus (**Figure 1**) is very simple.

One interesting method is a combination of solution dispersion and spray drying. The dispersion containing the nanoparticles in the polymer solution, after some time for homogenization, is injected in the spray dryer, which dries the material into a powder. The major drawback of this technique is the final yield of the material. However, this procedure occurs in just one step after all conditions are adjusted and the final material presents good dispersion and distribution. In this process, drugs can be injected together with nanosystems for controlled or targeted delivery. The equipment used in this process is shown in **Figure 2**.

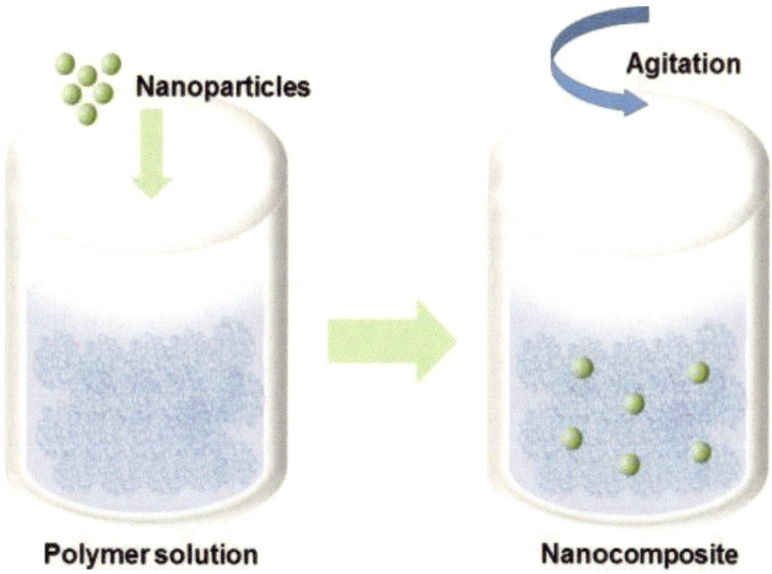

Figure 1. Scheme of obtaining polymer nanocomposites by the solution method.

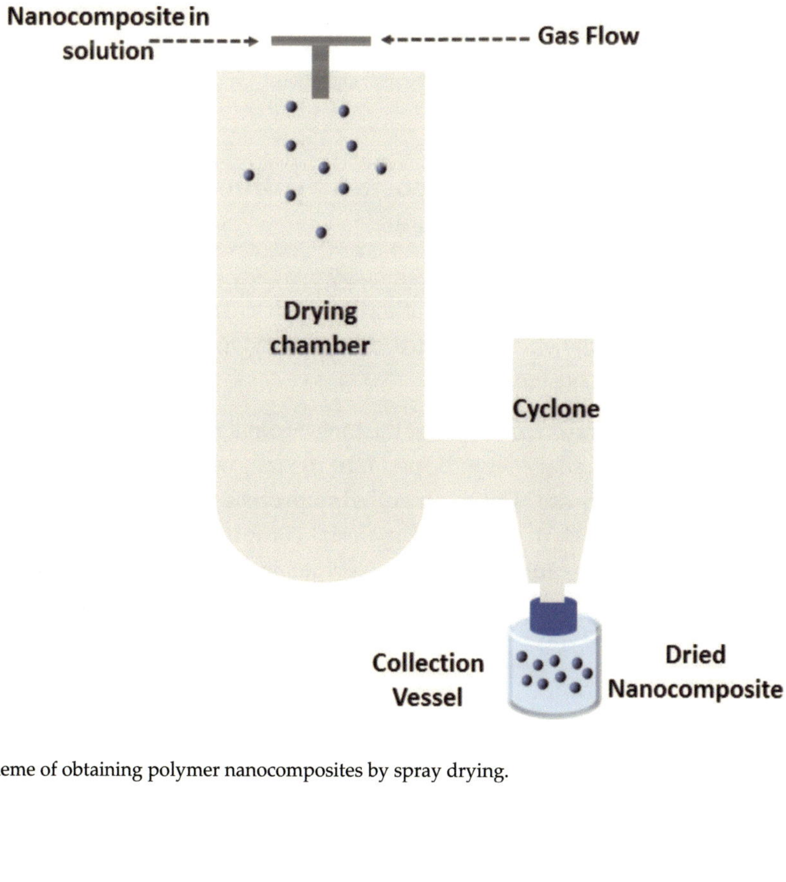

Figure 2. Scheme of obtaining polymer nanocomposites by spray drying.

2.3. Melt extrusion

This method has a major advantage in relation to the others since no solvent is necessary. However, the quantity of nanoparticles to be dispersed is very important. This method requires close monitoring of the nanoparticles' dispersion, because these agglomerate easier than in other methods. The apparatus is the same used for polymer processing without nanoparticles. Therefore, the researcher needs to pay attention to the temperatures used so as not to degrade the polymer during extrusion and also must pay attention to the time necessary for the nanoparticles to disperse properly. For natural polymers and few biopolymers, the degradation and melting temperatures are very close [7, 12, 17]. A typical extruder is shown in **Figure 3**.

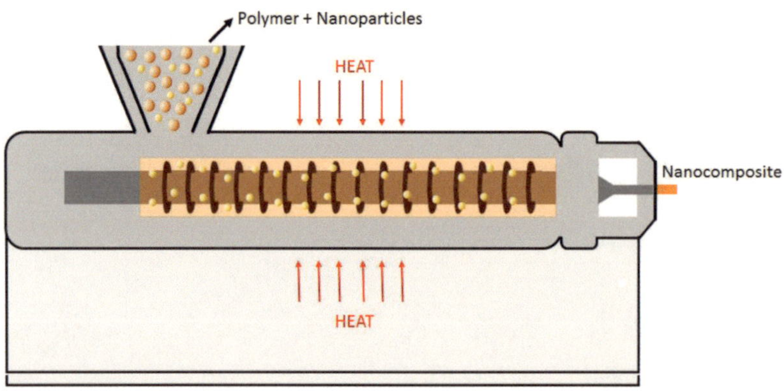

Figure 3. Scheme of obtaining polymer nanocomposites by melt extrusion.

3. Examples of obtaining nanocomposites by different methods and determination of their characteristics

This section gives examples of the three methods and discusses a new technique for their characterization: nuclear magnetic resonance relaxometry to measure proton spin-lattice relaxation time, with a time constant of T_1H.

The spin-lattice relaxation time is the time that the longitudinal magnetization takes to recover about 63% of its initial value after being flipped into the magnetic transverse plane by a perpendicular radio-frequency pulse. This is an enthalpy process where the energy of the spins is passed to the whole network to which the material conforms. Thus, T_1H can indicate the change in molecular mobility caused by intermolecular interactions and the distribution of the components of the material. Longer times refer to more rigid domains and shorter times indicate greater mobility [4–8, 18].

3.1. *In situ* polymerization

Obtaining poly-vinyl alcohol (PVA) and clay systems and use of relaxometry: The nanomaterials were prepared by *in situ* polymerization employing different clay ratios, from 0.25 to 10%

http://dx.doi.org/10.5772/intechopen.68142

w/w. Solutions of vinyl acetate were poured into Erlenmeyer flasks containing the clay dispersions. After 2 hours of heating at 60–65°C, 5 ml of benzoyl peroxide solution in methanol was add to acetate vinyl solution and the polymerization reaction occurred during 24 hours.

The main characterization was performed by measuring T_1H, with the following conditions. The Free Induction Decay (FID) amplitude was 40 points, ranging from 0.1 to 5000 ms, with four measurements for each point and a recycle interval of 5 s. **Table 1** reports the T_1H values for PVA and its nanocomposites with clay obtained by *in situ* polymerization.

Analysis of the T_1H data for the PVA-clay systems of the material containing 3% clay showed a significant decrease in the relaxation parameter, indicating good interaction between clay lamellae and the polymer matrix, in turn promoting good dispersion and distribution of the nanoparticles in the polymer matrix. The nanocomposite formed was exfoliated since the polymer's main chains were free to move around the clay lamella. This can also be explained by the presence of a paramagnetic metal in the clay structure, which acts as a relaxing agent. After this proportion, a tendency to increase the relaxation value was observed, showing that these proportions produce a predominance of intercalated chains, because they are not free to move since they are constrained among the clay lamellae.

3.2. Solution method

3.2.1. Obtaining PVA-clay films by solution casting

The films obtained by solution casting were prepared mixing a polymer solution and clay dispersion in water. After homogenization of both the solution and dispersion, the solvent was eliminated by heating in oven at 80°C. The films were also analyzed by measuring the proton spin-lattice relaxation time (**Table 2**) employing the same conditions as for the *in situ* polymerization.

According to the relaxation data in **Table 2**, the same behavior was found for the samples generated by *in situ* polymerization. Therefore, the effect was more pronounced and the samples obtained by solution intercalation showed better exfoliation for 3 and 5% clay in the PVA matrix. For 5% clay, the dispersion and distribution of the clay lamellae in the polymer matrix were very good and a predominance of exfoliated nanocomposite was obtained for both proportions, although at 5% this was more accentuated.

Sample/PVA-clay	T_1H (ms)
PVA	115
PVA-0.25	110
PVA-0.75	120
PVA-1.5	105
PVA-3.0	100
PVA-5.0	105
PVA-7.0	115

Table 1. T_1H values for PVA and its nanocomposites obtained by *in situ* polymerization.

Sample/PVA-clay	T_1H (ms)
PVA	120
PVA-0.25	115
PVA-0.75	105
PVA-1.5	105
PVA-3.0	95
PVA-5.0	70
PVA-7.0	90

Table 2. T_1H values for PVA and its nanocomposites obtained by solution casting.

3.2.2. Obtaining PC-organoclay films by solution casting

The PC films and PC-organoclay were obtained using chloroform as solvent and mixing the polymer solution and organoclay dispersion. The films were dried at room temperature under air circulation and the materials obtained were evaluated by measuring the T_1H values. The results are listed in **Table 3**.

Evaluating the T_1H Values, one can see that the incorporation of the organoclay in the polycarbonate through intercalation by solution promoted the generation of mixed nanomaterials, that is, with an exfoliated part and intercalated part, since there was no significant decrease in the spin-lattice relaxation time of the hydrogen nucleus, except at the proportion of 5%, which presented a higher degree of exfoliation. This can be explained because when the polymer chains are around the clay lamellae, the metals present in the clay composition (such as iron, calcium and magnesium) act as relaxing agents, causing a decrease in the relaxation times.

3.2.3. Obtaining PC-TiO_2 films by solution casting

Films of PC and PC-TiO_2 were obtained using chloroform as solvent by mixing the polymer solution and TiO_2 dispersion. The films were dried at room temperature with air circulation and the materials obtained were evaluated by measuring the T_1H values. The results are listed in **Table 4**.

Sample/PC-organoclay (%)	T_1H (ms)
PC	100
PC-1	105
PC-3	90
PC-5	95

Table 3. T_1H values of PC films and PC-organoclay obtained through solution casting.

Sample /PC-TiO_2 (%)	T_1H (ms)
PC	100
PC-1	95
PC-3	98
PC-5	90

Table 4. T_1H measurements of PC and PC-TiO_2 films.

The addition of titanium dioxide caused a greater decrease in the spin-lattice relaxation time of the hydrogen nucleus, mainly at the 5% ratio, which can be related to good dispersion of this oxide in the polymer matrix.

3.3. Melt extrusion

3.3.1. Obtaining PC-organoclay systems by extrusion

The materials were processed in a twin-screw extruder. The polymer and the organoclay (PC/organoclay) were mixed together at 270°C. The relaxometry characterization was done by measurement of T_1H (**Table 5**).

In the melt intercalation at 270°C, the systems containing 3 and 5% organoclay showed formation of hybrid nanomaterials with some tendency of exfoliation. This system, when compared to the same system obtained by solution, presented higher intercalation.

3.3.2. Obtaining PC-TiO_2 systems by extrusion

The PC/TiO_2 systems were processed in a twin-screw extruder at 270°C. The relaxometry characterization was done by measuring T_1H. The results are reported in **Table 6**.

The addition of titanium dioxide did not interfere in the relaxation time of the polymer matrix, so the values are similar to those for PC. This behavior may indicate that the particles were not well dispersed in the polymer matrix, probably because the processing temperature was not high enough to promote good dispersion of the particles in the polymer matrix.

Sample/PC-organolay (%)	T_1H (ms)
PC	95
PC-1	95
PC-3	90
PC-5	90

Table 5. T_1H measurements of the PC-organoclay nanocomposites obtained by melt intercalation.

Sample/PC-TiO_2 (%)	T_1H (ms)
PC	95
PC-1	93
PC-3	95
PC-5	93

Table 6. T_1H measurements of the nanocomposites formed by PC/TiO_2 obtained by melt intercalation at 270°C.

4. Some applications of nanocomposites

Nanocomposite materials can be applied in many areas, such as manufacture of plastic articles, medical devices, drug delivery systems, dental materials, military materials, sensors, automotive parts, biodegradable materials and others. In the health area, the use of these materials has been gaining prominence in recent years. The application of nanotechnology enables earlier diagnosis and better treatment of some diseases. The principal application of this technology involves controlled drug release and restoration or regeneration of tissues (epithelial, bone and dental tissues) [19–22].

4.1. Dental restoration

The incorporation of nanoparticles in dental composites has led to important improvements in clinical practice. The application of nanoparticles in these materials can improve mechanical properties, surface smoothness and gloss and reduce the polymerization shrinkage [23–25].

Mechanically speaking, the incorporation of nanoparticles in dental composites can improve several properties, such as wear resistance, elastic modulus, flexural strength, diametral tensile strength and fracture toughness. The use of nanotechnology in polymeric materials includes the use of many nanofiller types and several studies have shown the ability of these fillers to increase the hardness and decrease the roughness of the final restoration.

The use of nanocomposites in dental restoration aims to mimic the natural composition of the tooth tissues. The human teeth are formed mainly by two highly mineralized tissues, dentin and enamel (**Figure 4**). These tissues have an inorganic portion composed mainly of natural calcium-based nanoparticles such as hydroxyapatites and calcosperites.

The enamel is formed mostly (96–99%) of hydroxyapatite nanocrystals. Dentin, on the other hand, has a more complex composition, presenting lower percentage of nanohydroxyapatite and calcospherites, which compose about 70% of the tissue. They are dispersed between the organic matrix, which is composed mainly of extracellular matrix and collagen fibrils.

The following example provides a better understanding of the role of nanoparticles in the properties of dental restorative materials.

Figure 4. Dental tissue structure.

4.1.1. Dental resin with nanoparticles

To obtain these systems, a standardized resin matrix was used containing a mixture of bisphenol A glycidyl methacrylate (Bis-GMA), urethane dimethacrylate (UDMA) and triethylene glycol dimethacrylate (TEGDMA) in the proportion 35:40:25wt%, associated with a cam phorquinone as photoinitiator. As fillers were organo-modified clay and organo-modified silica. These nanoparticles presented diameters between 100 and 200 nm and thickness of 1 nm for the clay and a diameter of 16 nm for the silica.

In all groups, the proportion of nanofiller in the matrix was 2.5, 5 or 10% wt/wt to evaluate the influence of nanofiller in these systems. The resulting materials were compared with a resin with alumino-silicate microfillers (4 μm) [26–32]. The filler and matrix were mixed using a centrifugal mixing device. After obtaining these systems, they were evaluated by measuring the microhardness of each one.

In the Knoop hardness test (**Figure 5**), the best results were attained for the resins containing nanoparticles for all concentrations evaluated. Among the nanoparticle concentrations, the most promising results were observed in the groups containing 2.5% nanoclay and in groups containing 10% silica.

The addition of these fillers increases mechanical properties because they have a higher relative surface area than micro-particles. In this case, when the nanofiller presents good intermolecular interaction with the polymer matrix, it can be better dispersed and distributed, producing a larger polymer/nanofiller interfacial area and facilitating the transmission of forces between them.

The resin's properties need to mimic as closely as possible the physical and mechanical characteristics of dentin and enamel, so that the restorations can resist occlusal stresses. Materials containing nanoparticles are a promising alternative since the particles promote a greater durability and longevity of restorations.

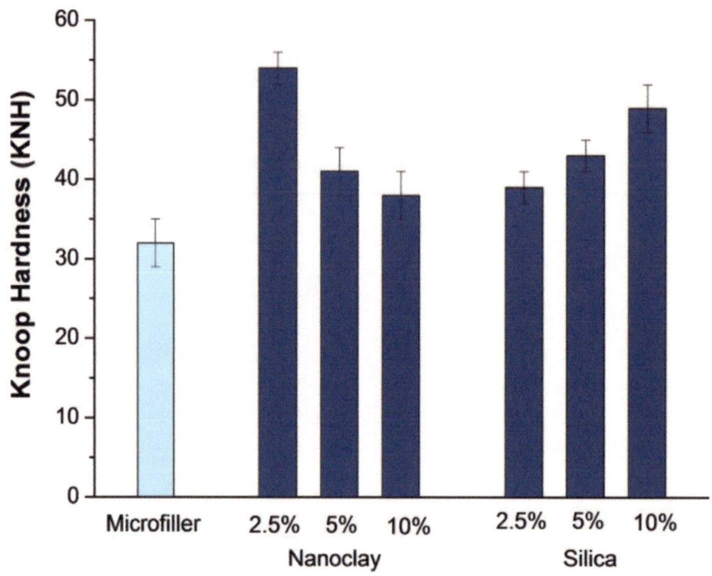

Figure 5. Knoop hardness results of dental resins with clay and silica nanoparticles.

4.2. Drug delivery systems

Polymeric nanoparticles or nanocomposites have promising features for drug delivery systems, including longer drug circulation time, better targeting to a specific tissue and reduced toxicity and adverse events, making disease treatment more tolerable to patients [19, 21].

In this context, nanocomposites using clay particles have drawn interest in recent years due to the significant changes promoted by small amounts of clay. These loads have several applications, including in polymer matrices aiming at improving mechanical properties, as also seen for dental restorations. To better understand the role of clay nanoparticles in drug delivery systems, the following example can be cited.

4.2.1. Chlorhexidine/nanoclay adhesive systems

To obtain these systems a standardized adhesive matrix composed of a dimethacrylate copolymer was used. The drug chlorhexidine diacetate was added to this material in the presence or absence of nanoclay particles. The nanofiller was used at 0.2% wt/wt in relation to the matrix, and the drug was tested in two different concentrations (0.5 and 1.0% wt/wt).

The *in vitro* drug release profiles of the systems containing chlorhexidine and doxycycline in the presence or absence of clay nanoparticles were analyzed and the results are shown in **Figure 6**.

The systems containing clay allowed prolonged drug release. This behavior occurs due to the lamellar conformation within the polymeric matrix. The clay layers interfere in the preferential diffusion pathways of the drug, creating more tortuous pathways, delaying the drug diffusion (**Figure 7**). Thus, after the release of the more external drug molecules, the more internal ones will be gradually released.

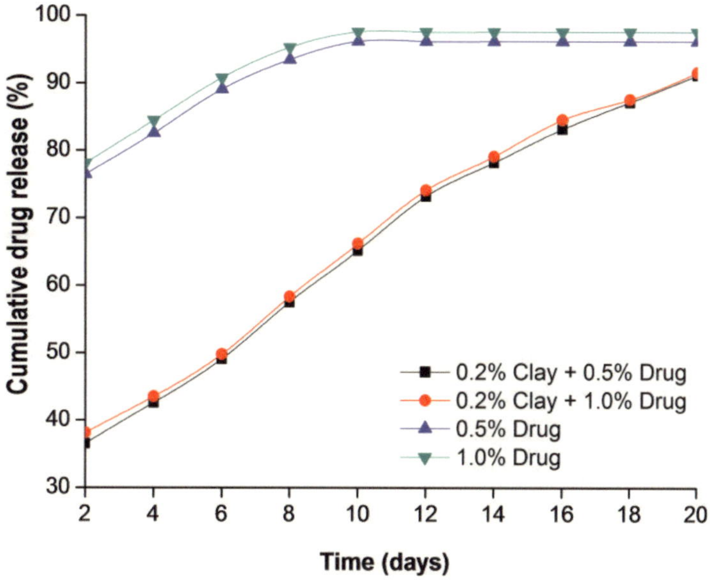

Figure 6. Drug release of the systems containing clay.

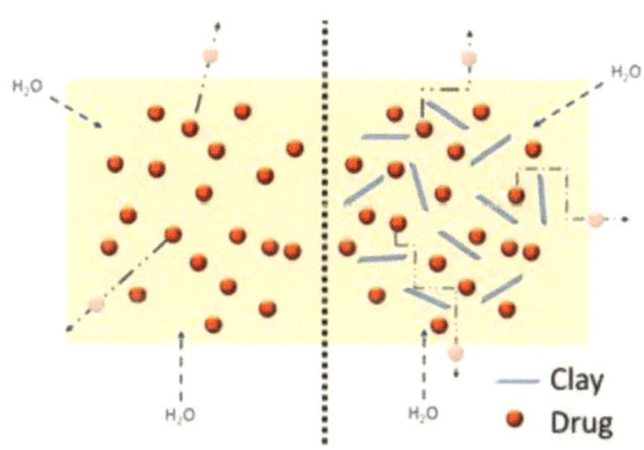

Figure 7. Schematic representation of drug release in the absence and presence of clay lamellae.

5. Polymer nanocomposite degradation

The statement that "polymers do not degrade" is not correct. Polymers are a large group of macromolecules and this group can be split into some categories: natural polymers, biopolymers, synthetic polymers and elastomers. Biopolymers without treatment degrade faster than synthetic polymers. This characteristic makes them attractive as substitutes for many products. The main degradation agent is oxygen, but it is not alone. Light, stress and bio-organisms can have the same effect. Each kind of degradation has a specific agent or combination of agents. In this context, the following causes of degradation are highlighted:

(a) Photodegradation: This is the degradation process that happens when the polymer molecules absorb photons from sunlight [33–36].

(b) Oxy-degradation: This degradation is caused by the action of oxygen on the materials, as happens to metals. In practice, it is easy to confirm this degradation because it can promote the appearance of chromophore groups that change the material's color [33, 35].

(c) Photo-oxy-degradation: This degradation process is a combination of the two previous ones. In some polymers, after the light radiation creates free radicals the oxygen reacts with them, generating carbonyl groups [33, 35].

(d) Biodegradation: It is caused by the action of enzymes produced by microorganisms in materials susceptible at enzymatic rupture [37].

(e) Thermal degradation: High temperature gives molecules energy, which vibrates and deforms them, but it also can break the chain bonds, prompting the loss of properties. It depends on the chemical structure and molar mass [38].

(f) Chemical degradation: Each polymer has its own chemical structure and this structure is susceptible to chemical attack that can change or break the molecular structure due to the interaction with chemical compounds that are stronger than the chemical bond [39].

Nanoparticles have different properties. Their small size changes many properties of materials, even in small amounts. The interactions between matrix and nanoparticles can promote interesting results.

Polymer nanocomposites often present synergism between the polymer and nanoparticles, which can change how the matrix degrades. This characteristic can extend or shorten the life of the biopolymer matrix, enabling tailoring materials for specific uses. For instance, faster degradation reduces the pollution generated by these materials [39].

5.1. Clay polymer nanocomposites

Various types of clay are widely used because they are versatile and can cause chemical modification of the matrix by changing the exchangeable cations. The small size of clay nanoparticles promotes better compatibility between clay and polymer matrix. This helps to disperse the clay in the matrix, which can reduce photo-oxy-degradation by acting as a physical barrier to oxygen and reflecting UV light because of the particles' large aspect ratio. Also, they can reduce the effect of thermal and chemical degradation due to stronger interactions between clay and matrix.

5.2. Silver nanocomposites

Silver in nanoscale has bactericidal properties and provides UV light protection. This former happens because in nanoscale the asepsis of the oligodynamic effect is improved, killing bacteria. The latter effect is due to reflection of light, blocking photodegradation.

5.3. Silica nanocomposites

Silica is hydrophilic due to the presence of Si-OH. It can be used to speed up the degradation of synthetic polymer matrixes by microorganisms that need water. Alternatively, the surface can be modified to change it from hydrophilic to hydrophobic, thus improving the resistance of biopolymers against microorganisms.

5.4. Carbon nanotube nanocomposites

Carbon nanotubes can have a single wall (CNTs) or multiple walls (MWCNTs). These nanoparticles are commonly used to change polymers that are nonconductors to semi-electrical conductors. Also, CNTs improve the resistance of polymers to UV degradation, mechanical stress and chemical degradation as well as reducing thermal stress.

6. Final Comments

The focus of this chapter has been on providing a basic understanding of polymer nanocomposites, especially methods to obtain them, with some examples of each one, and comparing the results obtained by each method for the same polymer system and the effects of degradation in the presence of each nanoparticle cited. It also covers low-field NMR relaxometry, a characterization tool that can efficiently evaluate nanoparticles' dispersion and distribution in the polymer matrix, according to the preparation method.

Author details

Maria Inês Bruno Tavares*, Emerson Oliveira da Silva, Paulo Rangel Cruz da Silva and Lívia Rodrigues de Menezes

*Address all correspondence to: mibt@ima.ufrj.br

Federal University of Rio de Janeiro, Institute of Macromolecules Professor Eloisa Mano, IMA/UFRJ, Rio de Janeiro, Brazil

References

[1] Ashwani Sharma, Pallavi Sanjay Kumar. Synthesis and Characterization of CeO-ZnO Nanocomposites. Nanoscience and Nanotechnology. 2012;**2**(3):82-85. DOI: 10.5923/j.nn.20120203.07

[2] Swetha Chandrasekaran, Gabriella Faiella, L.A.S.A. Prado, Folke Tölle, Rolf Mülhaupt, Karl Schulte. Thermally reduced graphene oxide acting as a trap for multiwall carbon

nanotubes in bi-filler epoxy composites. Composites Part A: Applied Science and Manufacturing. 2013;**49**:51-57. DOI: 10.1016/j.compositesa.2013.02.008

[3] Nicolas Jacquel, Chi-Wei Lo, Ho-Shing Wu, Yu-Hong Wei, Shaw S Wang. Solubility of polyhydroxyalkanoates by experiment and thermodynamic correlations. AIChE journal. 2007;**53**(10):2704-2714. DOI: 10.1002/aic.11274

[4] Mariana Bruno Rocha e Silva, Maria Inês Bruno Tavares, Emerson Oliveira da Silva, Roberto Pinto Cucinelli Neto. Dynamic and structural evaluation of poly (3- hydroxybutyrate) layered nanocomposites. Polymer Testing. 2013;**32**(1):165-174. DOI: 10.1016/j.polymertesting.2012.09.006

[5] Maria Ines Bruno Tavares, Regina Freitas Nogueira, Rosane Aguiar da Silva San Gil, Monica Preto, Emerson Oliveira da Silva, Mariana Bruno Rocha e Silva, Eduardo Miguez. Polypropylene–clay nanocomposite structure probed by H NMR relaxometry. Polymer Testing. 2007;**26**(8):1100-1102. DOI: 10.1016/j.polymertesting.2007.07.012

[6] DL VanderHart, A Asano, JW Gilman. Solid-state NMR investigation of paramagnetic nylon-6 clay nanocomposites. 1. Crystallinity, morphology, and the direct influence of Fe3+ on nuclear spins. Chemistry of Materials. 2001;**13**(10):3781-3795. DOI: 10.1021/cm0110775

[7] Paulo S. R. C. da Silva, Maria I. B. Tavares. Intercalação por Solugão de Poliestireno de. 2013;**23**(5):644-648. DOI: 10.4322/polimeros.2013.047

[8] SILVA, Paulo Sergio Rangel Cruz da; TAVARES, Maria Inês Bruno. Solvent Effect on the Morphology of Lamellar Nanocomposites Based on HIPS. Materials Research. 2015;18(1):191-195. DOI: 10.1590/1516-1439.307314

[9] Fernanda Abbate dos Santos, Maria Inês Bruno Tavares. Development of biopolymer/cellulose/silica nanostructured hybrid materials and their characterization by NMR relaxometry. Polymer Testing. 2015;47:92-100. DOI: 10.1016/j.polymertesting.2015.08.008

[10] Guido Kickelbick, editor. Hybrid Materials: Synthesis, Characterization, and Applications. Federal Republic of German: John Wiley & Sons; 2007. 516 p.

[11] Monteiro, Mariana S. S. B.; Rodrigues, Claudia Lopes; Neto, Roberto P. C.; Tavares, Maria Inês Bruno. The Structure of Polycaprolactone-Clay Nanocomposites Investigated by 1H NMR Relaxometry. Journal of Nanoscience and Nanotechnology. 2012;**12**(9):7307-7313. DOI: 10.1166/jnn.2012.6431

[12] Ana Claudia S Valentim, Maria Inês Bruno Tavares, Emerson Oliveira Da Silva. The effect of the Nb2O5 dispersion on ethylene vinyl acetate to obtain ethylene vinyl acetate/Nb2O5 nanostructured materials. Journal of nanoscience and nanotechnology. 2013;**13**(6):4427-4432. DOI: 10.1166/jnn.2013.7162

[13] Soares, Igor Lopes; Chimanowsky, Jorge Pereira; Luetkmeyer, Leandro; Silva, Emerson Oliveira da; Souza, Diego de Holanda Saboya; Tavares, Maria Inês Bruno. Evaluation of the Influence of Modified TiO2 Particles on Polypropylene Composites. Journal of Nanoscience and Nanotechnology. 2015;**15**(8):5723-5732. DOI: 10.1166/jnn.2015.10041

[14] Alessandra dos Santos Almeida, Maria Inês Bruno Tavares , Emerson Oliveira da Silva, Roberto Pinto Cucinelli Neto, Leonardo Augusto Moreira. Development of hybrid nanocomposites based on PLLA and low-field NMR characterization. Polymer Testing. 2012;**31**(2):267-275. DOI: 10.1016/j.polymertesting.2011.11.005

[15] Mariana Sato de Souza de Bustamante Monteiro; Roberto Pinto Cucinelli Neto; Izabel Cristina Souza Santos; Emerson Oliveira da Silva; Maria Inês Bruno Tavares. Inorganic-organic hybrids based on poly (e-Caprolactone) and silica oxide and characterization by relaxometry applying low-field NMR. Materials Research. 2012;**15**(6):825-832. DOI: 10.1590/S1516-14392012005000121

[16] Antonio de Pádua C B Cunha, Maria Inês Bruno Tavares, Emerson Oliveira Silva, Soraia Zaioncz. The Effect of Montmorillonite Clay on the Crystallinity of Poly(vinyl alcohol) Nanocomposites Obtained by Solution Intercalation and In Situ Polymerization. Journal of Nanoscience and Nanotechnology . 2015;**15**(4):2814-2820. DOI: 10.1166/jnn.2015.9233

[17] M.R. TavaresL.R. de MenezesD.F. do NascimentoD.H.S. SouzaF. ReynaudM.F.V. MarquesM.I.B. Tavares. Polymeric nanoparticles assembled with microfluidics for drug delivery across the blood-brain barrier. The European Physical Journal Special Topics. 2016;**225**(4):779-795. DOI: 10.1140/epjst/e2015-50266-2

[18] Jorge P. Chimanowsky Junior, Igor Lopes Soares, Leandro Luetkmeyer, Maria Inês Bruno Tavares. Preparation of high-impact polystyrene nanocomposites with organo-clay by melt intercalation and characterization by low-field nuclear magnetic resonance. Chemical Engineering and Processing: Process Intensification. 2014;**77**:66-76. DOI: 10.1016/j.cep.2013.11.012

[19] S. Olejniczak, S. Kazmierski, P. K. Pallathadka, M. J. Potrzebowski. A review on advances of high-resolution solid state NMR spectroscopy. Polymery. 2007;**10**:713-721.

[20] Audumbar Digambar Mali, Ritesh Bathe, Manojkumar Patil. An updated review on transdermal drug delivery systems. International Journal of Advances in Scientific Research. 2015;**1**(6):244-254. DOI: 10.7439/ijasr.v1i6.2243.

[21] Toral Patel, Jiangbing Zhou, Joseph M. Piepmeiera, W. Mark Saltzman. Polymeric nanoparticles for drug delivery to the central nervous system. Polymeric nanoparticles for drug delivery to the central nervous system. 2012;**67**(7):701-705. DOI: 10.1016/j.addr.2011.12.006

[22] Vincent J. McBrierty, Kenneth J. Packer. Nuclear Magnetic Resonance in Solid Polymers (Cambridge Solid State Science Series). United State: Cambridge University Press; 1993. **351** p. DOI: 10.1017/CBO9780511525278

[23] Stefanie Wohlfart, Svetlana Gelperina, Jörg Kreuter. Transport of drugs across the blood–brain barrier by nanoparticles. Journal of Controlled Release. 2012;**161**(2):264-273. DOI: 10.1016/j.jconreL2011.08.017

[24] Elise Lepeltier, Claudie Bourgaux, Patrick Couvreur. Nanoprecipitation and the "Ouzo effect": Application to drug delivery devices. Advanced Drug Delivery Reviews. 2014;**71**:86-97. DOI: 10.1016/j.addr.2013.12.009

[25] Elvin Blanco, Haifa Shen, Mauro Ferrari. Principles of nanoparticle design for overcoming biological barriers to drug delivery. Nature Biotechnology. 2015;**33**(9):941- 951. DOI: 10.1038/nbt.3330

[26] Sébastien Beun, Thérèse Glorieux, Jacques Devaux, José Vreven, Gaëtane Leloup. Characterization of nanofilled compared to universal and microfilled composites. Dental Materials. 2007;**23**(1):51-59. DOI: 10.1016/j.dental.2005.12.003

[27] Jirun Sun, Aaron M. Forster, Peter M. Johnson, Naomi Eidelman, George Quinn, Gary Schumacher, Xinran Zhang, Wen-li Wu. Improving performance of dental resins by adding titanium dioxide nanoparticles. Dental Materials. 2011;**27**(10):972-982. DOI: 10.1016/j.dental.2011.06.003

[28] E. Helal, N.R. Demarquette, L.G. Amurin, E. David, D.J. Carastan, M. Fréchette. Styrenic block copolymer-based nanocomposites: Implications of nanostructuration and nanofiller tailored dispersion on the dielectric properties. Polymer. 2015;**64**:139-152. DOI: 10.1016/j.polymer.2015.03.026

[29] Yasser Zare. Estimation of material and interfacial/interphase properties in clay/polymer nanocomposites by yield strength data. Applied Clay Science. 2015;**115**:61-66. DOI: 10.1016/j.clay.2015.07.021

[30] J. Silvestre, N. Silvestre, J. de Brito. An Overview on the Improvement of Mechanical. Journal of Nanomaterials. 2015;2015:1-13. DOI: 10.1155/2015/106494

[31] Yasser Zare. Study on interfacial properties in polymer blend ternary nanocomposites: Role of nanofiller content. Computational Materials Science. 2016;**111**:334-338. DOI: 10.1016/j.commatsci.2015.09.053

[32] Sébastien Beun, Christian Bailly, Jacques Devaux, Gaëtane Leloup . Physical, mechanical and rheological characterization of resin-based pit and fissure sealants compared to flowable resin composites.. Dental Materials. 2012;**28**(4):349-359. DOI: 10.1016/j.dental.2011.11.001

[33] A. Rivera, L. Valdés, J. Jiménez, I. Pérez, A. Lam, E. Altshuler, L.C. de Ménorval, J.O. Fossum, E.L. Hansen, Z. Rozynek. Smectite as ciprofloxacin delivery system: Intercalation and. Applied Clay Science. 2016;**124**:150-156. DOI: 10.1016/j.clay.2016.02.006

[34] Keleher J, Thomas C, Kevin J, Bianca G, Mlynarski A, Brain S, et al.. Synthesis and Characterization of a Chitosan/PVA Antimicrobial Hydrogel Nanocomposite for Responsive Wound Management Materials. Journal of Microbial & Biochemical Technology. 2016;**8**:065-070. DOI: 10.4172/1948-5948.1000264

[35] Paulo Sergio Rangel Cruz da Silva, Maria Inês Bruno Tavares. The use of relaxometry to evaluate the aging process in hybrid HIPS nanocomposites. Polymer Testing. 2015;**48**:115-119. DOI: 10.1016/j.polymertesting.2015.10.005

[36] Morais, Frederick Louis Dias de; Medeiros, Felipe da Silva; Silva, Glaura Goulart; Rabello, Marcelo Silveira; Sousa, Alexandre Rangel de. Photodegradation of UHMWPE Filled with Iron Ore Fine. Materials Research. 2017;**20**(2): 356-364. DOI: 10.1590/1980-5373-mr-2016-0320

[37] Santos, Fernanda Abbate dos, & Tavares, Maria Inês Bruno. Development and characterization of hybrid materials based on biodegradable PLA matrix, microcrystalline cellulose and organophilic silica. Polímeros. 2014;**24**(5):561-566. DOI: 10.1590/0104-1428.1653

[38] Jorge P. Chimanowsky Jr., Roberto Pinto Cucinelli Neto, Maria Inês Bruno Tavares. NMR evaluation of polystyrene nanocomposites degraded by repeated extrusion processing. Polymer Degradation and Stability. 2015;**118**:178-187. DOI: 10.1016/j.polymdegradstab.2015.03.022

[39] Harwood, H. J, editors. Polymer Degradation and Stabilization. 1 edition st ed. Springer; 1984. 122 p.

Chapter 8

Synthesis Characterization of Nanostructured $ZnCo_2O_4$ with High Sensitivity to CO Gas

Juan Pablo Morán-Lázaro, Florentino López-Urías,
Emilio Muñoz-Sandoval, Oscar Blanco-Alonso,
Marciano Sanchez-Tizapa,
Alejandra Carreon-Alvarez, Héctor Guillén-Bonilla,
María de la Luz Olvera-Amador,
Alex Guillén-Bonilla and
Verónica María Rodríguez-Betancourtt

Additional information is available at the end of the chapter

http://dx.doi.org/10.5772/68043

Abstract

In this work, nanostructured $ZnCo_2O_4$ was synthesized via a microwave-assisted colloidal method, and its application as gas sensor for the detection of CO was studied. Typical diffraction peaks corresponding to the cubic $ZnCo_2O_4$ spinel structure were identified at calcination temperature of 500°C by X-ray powder diffraction. A high degree of porosity in the surface of the nanostructured powder of $ZnCo_2O_4$ was observed by scanning electron microscopy and transmission electron microscopy, faceted nanoparticles with a pockmarked structure were clearly identified. The estimated average particle size was approximately 75 nm. The formation of $ZnCo_2O_4$ material was also confirmed by Raman characterization. Pellets fabricated with nanostructured powder of $ZnCo_2O_4$ were tested as sensors using CO gas at different concentrations and temperatures. A high sensitivity value of 305–300 ppm of CO was measured at 300°C, indicating that nanostructured $ZnCo_2O_4$ had a high performance in the detection of CO.

Keywords: spinel, nanoparticles, cobaltite, sensors, characterization, synthesis

INTECH
open science | open minds

1. Introduction

Gas sensors technology has numerous applications in the automotive, industrial, domestic, and security sectors. In the automotive and industrial sectors, gas sensors are necessary to detect toxic and harmful gases for environment protection and human health (i.e., carbon monoxide). Sensor materials based on semiconducting metal oxides are one of the several technologies being used in the detection of pollutants [1]. This type of oxide materials is suitable for gas sensor applications due to their interesting structural, functional, physical, and chemical properties. To date, reports indicate that n-type semiconductor materials, such as SnO_2 [2, 3], ZnO [4], and TiO_2 [5] are most studied in gas sensing area. By contrast, a limited amount research works on p-type oxide semiconductor gas sensors have been found, the most studied being CuO, Co_3O_4, and NiO [6]. However, some sensor parameters such as gas sensitivity and working temperature still need to be improved. Therefore, additional studies are needed to improve the gas sensing characteristics of p-type semiconducting oxides by modifying factors such as synthesis conditions, structure, morphology, and composition.

Among the p-type semiconductors, zinc cobaltite ($ZnCo_2O_4$) is a material with spinel-type structure, which has been mainly used as electrode for Li-ion batteries [7–11] and supercapacitors [12–15], due to its higher electrochemical performances and higher conductivities. To date, sensor devices based on nanostructured spinel $ZnCo_2O_4$ have particularly exhibited an excellent sensitivity toward liquefied petroleum gas [16–18], ethanol [19, 20], acetone [21], Cl_2 [22], formaldehyde [23], and Xylene [24]. Additionally, several references [18–21, 25] reported its poor sensitivity to carbon monoxide (CO). On the other hand, a variety of synthesis methods have been developed on the preparation of $ZnCo_2O_4$ such as combustion [7], thermal decomposition [17], co-precipitation/digestion [18], W/O microemulsion [22], hydrothermal [9, 14], sol-gel [26], and surfactant-mediated method [27]. Recently, the colloidal route assisted by microwave radiation has provided an efficient and low cost synthesis method to obtain different types of nanostructured materials [28–31]. In this simple synthetic process, the addition of a surfactant agent plays a key role in the material's microstructure because the surfactant's ligands adsorb on the particles' surface inhibiting the particle growth and modifying the particles' microstructure [31–34]. Also, microwave radiation provides a rapid evaporation of the precursor solvent and a short reaction time in comparison with conventional heating [35, 36]. With this in mind, the synthesis of nanostructured $ZnCo_2O_4$ was done via a microwave-assisted colloidal method using zinc nitrate, cobalt nitrate, dodecylamine (as surfactant), and ethanol. Consequently, nanostructured $ZnCo_2O_4$ powder was characterized by X-ray powder diffraction (XRD), scanning electron microscopy (SEM), transmission electron microscopy (TEM), and Raman spectroscopy. The potential application of nanostructured $ZnCo_2O_4$ as gas sensor was studied by measuring its sensitivity toward different CO concentrations and working temperatures.

2. Synthesis, characterization, and gas sensing application of $ZnCo_2O_4$

2.1. Synthesis

For the preparation of nanostructured $ZnCo_2O_4$ by microwave-assisted colloidal method, first, 0.947 g of $Zn(NO_3)_2{\cdot}xH_2O$ (Zinc nitrate hydrate), 2.91 g of $Co(NO_3)_2{\cdot}6H_2O$ (cobalt nitrate hexahydrate), and 1 g of $C_{12}H_{27}N$ (dodecylamine) were dissolved separately in 5 mL of ethanol and kept under stirring for 20 min, at room temperature. Then, the cobalt nitrate solution was added drop wise to the dodecylamine solution under stirring. Then, the zinc nitrate solution was slowly added, producing a wine color solution with pH = 2. This resulting colloidal solution was stirred for approximately 24 h. Then, the solvent evaporation was made by a domestic microwave oven (General Electric JES769WK) operated at low power (~140 W). During the evaporation process, microwave radiation was applied for a period of 1 min, over a period of 3 h. The resulting solid material was dried in air at 200°C for 8 h using a muffle-type furnace (Novatech). Finally, the obtained powders were calcined at 500°C for 5 h. For each thermal treatment, a heating rate of 100°C/h was used. The resulting powders were black. The general synthesis process is illustrated in **Figure 1**.

2.2. Experimental techniques

The structural characterization was performed by XRD using a PANalytical Empyrean system (CuKα, λ = 1.546 Å). The XRD patterns were recorded, at room temperature, in the 2θ range of 10–70° using a step size of 0.02°. The morphological characterization was made by SEM, TEM, high-resolution transmission electron microscopy (HRTEM), energy-dispersive X-ray (EDS), and high-angle annular dark-field/scanning transmission electron microscopy (HAADF-STEM). For SEM studies, a FEI-Helios Nanolab 600 system operated at 20 kV was used, while a FEI Tecnai-F30 system operated at 300 kV was used for TEM, HRTEM, EDS, and HAADF-STEM analysis. The optical characterization was done by Raman spectroscopy using a Thermo Scientific DXR confocal Raman with a 633 nm excitation source. The Raman spectrum was recorded from 150 to 800 cm^{-1}, at room temperature, using a Laser power of 5 mW. The gas response (sensitivity) was acquired on pellets of $ZnCo_2O_4$ in the presence of several concentrations (1, 5, 50, 100, 200, and 300 ppm) of CO. The sensor devices were fabricated with 0.350 g of the nanostructured powder of $ZnCo_2O_4$, forming pellets with a thickness of 0.5 mm and a diameter of 12 mm. A TM20 Leybold detector was used to control the gas concentration and the partial pressure, and a digital-multimeter (Keithley) was put into use for the measurement of the electric resistance. A general schematic diagram of the gas sensing measurement system is shown in Ref. [33]. The sensitivity was defined as $S = R_a/R_g$, where R_a is the resistance measured in air and R_g is the resistances in gas [21, 24].

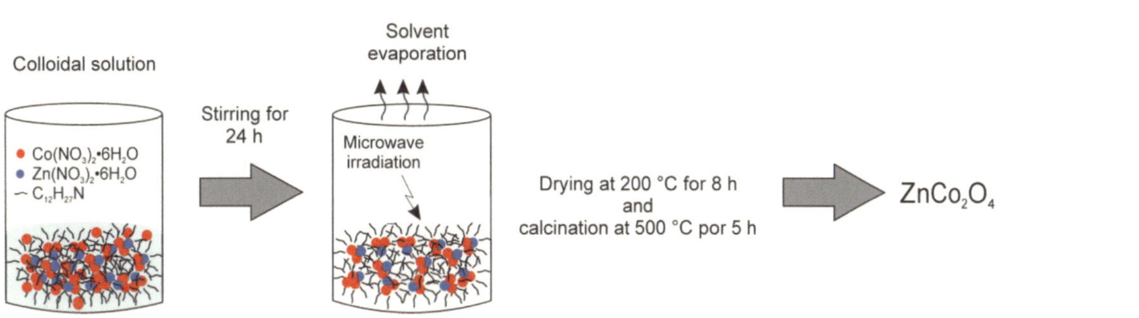

Figure 1. Schematic illustration of the synthesis of $ZnCo_2O_4$.

2.3. Structural analysis

Figure 2 shows a typical XRD pattern of the $ZnCo_2O_4$ powder obtained at a calcination temperature of 500°C. In this pattern, the main phase detected was the $ZnCo_2O_4$ crystal structure, which was identified by the JCPDF #23-1390 file. The sharp and strong peaks indicated a good crystallinity of the $ZnCo_2O_4$ sample. The diffraction peaks corresponded well to the (111), (220), (311), (222), (400), (422), (511), and (440) planes of the $ZnCo_2O_4$ spinel phase situated at 2θ = 18.95°, 31.21°, 36.80°, 38.48°, 44.73°, 55.57°, 59.28°, and 65.14°, respectively. The average crystal size, which was calculated by Scherrer's formula [38], using the XRD (311) plane, was ~24.7 nm. Since no secondary phase was detected in the XRD pattern of the $ZnCo_2O_{4,}$ the synthesis procedure developed in this work allowed to obtain the crystalline phase of $ZnCo_2O_4$ without additional diffraction peaks. Thus, this method of synthesis might also be useful for the preparation of other oxide materials.

2.4. Morphological investigations

Figure 3 shows the surface morphology of the $ZnCo_2O_4$ powder calcined at 500°C with different magnifications. **Figure 3a** exhibits a SEM image at low magnification, which revealed a surface with a high degree of porosity with pores of irregular shape. The average pore size was calculated around 724 nm. This extensive porosity has been associated with the emission of gas during the removal of organic matter in the calcination process [28]. At high magnification (**Figure 3b**), a large number of nanoparticles with irregular shape and size in the range of 50–110 nm were clearly observed. In colloidal chemistry, it is known

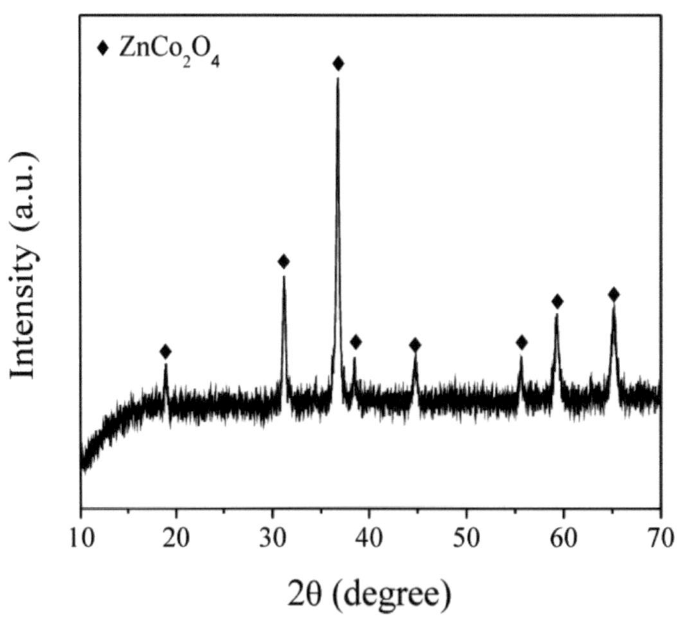

Figure 2. XRD pattern of the $ZnCo_2O_4$ powder after a thermal treatment at 500°C in air [37].

http://dx.doi.org/10.5772/68043

Figure 3. SEM images of the $ZnCo_2O_4$ powder at: (a) low (1,100x) and (b) high magnification (50,000x) [37].

that the formation of nanoparticles follows a nucleation and a growth mechanism [39]. In this mechanism, first, the nucleation is produced when the concentration of reagents reach the supersaturation limit for a short period of time. Consequently, the formation of large number of crystal nuclei occurs. Later, the process of growth of the particles is developed by diffusion. In our synthesis, the final solution does not present the supersaturation, although the nucleation process could occur when the zinc nitrate solution was added to the cobalt and dodecylamine solution, and the process of growth carried out is during the stirring of the final colloidal solution [31, 40]. On the other hand, the dodecylamine plays a key role in the microstructure of $ZnCo_2O_4$ particles, since the dodecylamine in the colloidal solution affects the particle growth via the saturating of nanocrystal surfaces and hence, results in the formation of $ZnCo_2O_4$ nanoparticles with a peculiar morphology (faceted nanoparticles were obtained) [28]. With thermal treatment at 500°C, the dodecylamine was finally removed from the $ZnCo_2O_4$ sample.

Figure 4 shows the typical TEM images of the $ZnCo_2O_4$ powder calcined at 500°C. **Figure 4a** exhibits a high concentration of nanoparticles, which was also observed by SEM. As can be seen **Figure 4b**, faceted nanoparticles with a pockmarked structure were clearly identified. The average particle size was ~75 nm, with a standard deviation of ±12.6 nm. A typical HRTEM image is displayed in **Figure 4c**. This image confirmed the presence of faceted nanoparticles and a value of 0.286 nm corresponded to the inter-planar d-spacing of the (200) plane of $ZnCo_2O_4$ spinel structure.

In order to investigate the nanoparticle's composition, EDS line scan was performed on $ZnCo_2O_4$ powder. **Figure 5** shows the corresponding analysis. **Figure 5a** shows a HAADF-STEM image of the $ZnCo_2O_4$ nanoparticles. The image confirms the presence of faceted nanoparticles with a pockmarked structure, which is consistent with the TEM images. In the EDS line scan, zinc, cobalt, and oxygen are observed across the linear mapping, confirming the presence of the expected elements, as seen in **Figure 5b**. In the central region X_2, a

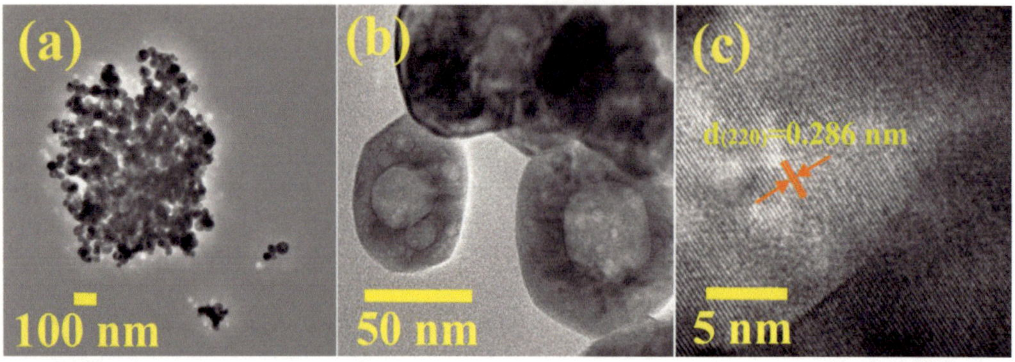

Figure 4. (a, b) TEM and (c) HRTEM images of $ZnCo_2O_4$.

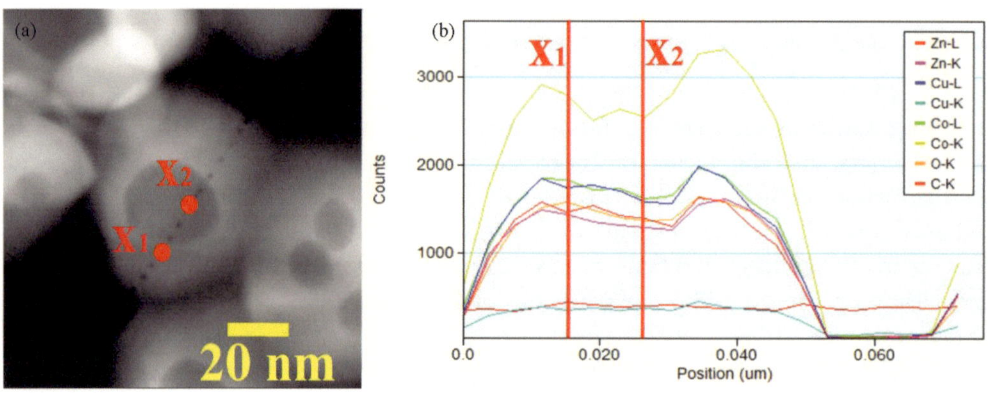

Figure 5. (a) HAADF-STEM image and (b) EDS elemental line scan of an individual $ZnCo_2O_4$ nanoparticle.

decrease of the element composition is observed in comparison to point X_1, which can be due to the irregular surface of the nanoparticle (pockmarked zone). It is also evident that cobalt exists in larger amount than zinc, corresponding to the target ratio of 1:2. However, the EDS line scan shows carbon (C) and copper (Cu) compounds, which are due to the sample support.

2.5. Raman characterization

The Raman spectrum shown in **Figure 6** allowed us to confirm the formation of the $ZnCo_2O_4$ when a calcination temperature of 500°C was used. As shown in **Figure 6**, the Raman spectrum of the $ZnCo_2O_4$ powder shows five vibrational bands located at 182, 475, 516, 613, and 693 cm^{-1} corresponding to the five active Raman bands of $ZnCo_2O_4$ spinel structure [41]. However, the band at 204 cm^{-1} is a vibrational mode that could be generated by Co_3O_4 [42]. The formation of this oxide is due to the cation disorder (substitution of Zn^{2+} by Co^{2+}) in

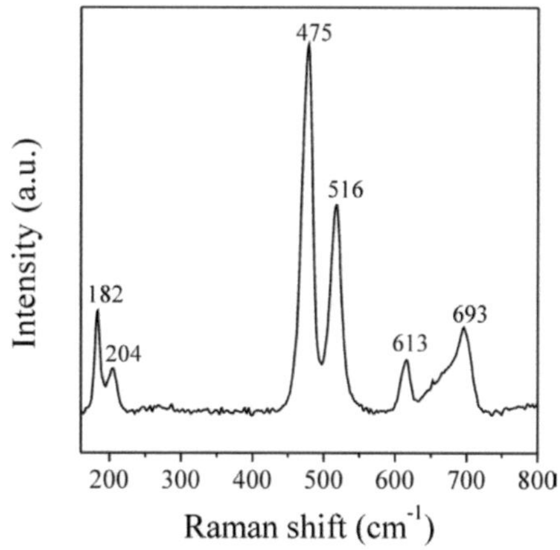

Figure 6. Raman spectrum of nanostructured $ZnCo_2O_4$ powder [37].

the $ZnCo_2O_4$ spinel structure. As the Co_3O_4 possess a spinel structure same as the $ZnCo_2O_4$; therefore, they have similar XRD patterns and a deformation in the XRD pattern of $ZnCo_2O_4$ is not expected.

2.6. Gas sensing application of $ZnCo_2O_4$

The sensing performance of the $ZnCo_2O_4$ sensor was investigated on pellets fabricated from the nanostructured $ZnCo_2O_4$ powders and tested in different concentrations of CO. **Figure 7** shows the variation of sensitivity against temperature at different concentrations of CO (1, 5, 50, 100, 200, and 300 ppm). As shown in **Figure 7**, only minor variations in sensitivity were measured at operating temperatures between 100 and 200°C in whole CO concentration range (1–300 ppm). For operating temperatures above 200°C, the sensitivity increased markedly from 5 to 300 ppm, with the maximum values of the sensitivity registered at 300°C.

Figure 8 shows sensitivity of the $ZnCo_2O_4$ sensor toward different concentrations of CO at 100, 200, and 300°C. As seen in **Figure 8a**, a sensitivity value of 1 was maintained across the CO concentration range when the $ZnCo_2O_4$ sensor was operated at 100°C. On the contrary, when the $ZnCo_2O_4$ sensor was working at a temperature of 200°C (**Figure 8a**) and 300°C (**Figure 8b**), the sensitivity increased with an increase of CO concentration. At 200°C, the sensitivity values of the sensor were 1, 1.1, 1.5, 2.2, 2.8, and 3.1 for CO concentrations of 1, 5, 50, 100, 200, and 300 ppm, respectively. However, at 300°C and with same concentrations, the sensitivity values were 1, 2, 3.3, 84.5, 157.5, and 305, respectively. The observed increase in sensitivity with the concentration is due to increase in gas concentration and operation temperature. The increase of the sensitivity is also associated with increased oxygen desorption at high temperatures [43, 44]. Additionally, the $ZnCo_2O_4$ sensor showed a decrease in gas response when CO gas were removed from the vacuum chamber.

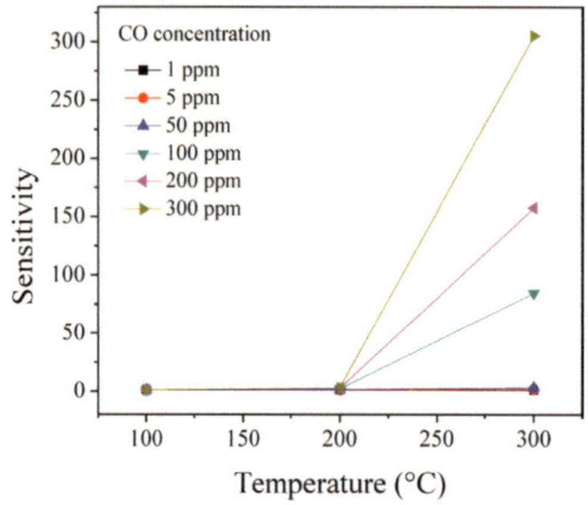

Figure 7. Sensitivity of the $ZnCo_2O_4$ sensor as a function of the temperature.

It is known that the gas sensing mechanism of semiconducting materials is based on the changes of the electrical resistance produced by interaction between the target gas and chemisorbed oxygen ions [45, 46]. When oxygen is adsorbed on the semiconductor's surface, oxygen species are generated at the surface by taking electrons from the conduction band of the semiconductor. In general, molecular (O_2^-) and ionic (O^- and O^{2-}) species are formed below 150°C and above this temperature, respectively [47]. Consequently, a space charge layer with thickness of ~100 nm is formed at the surface [6]. In our tests at temperatures above 100°C, the ionic species that adsorb chemically on the sensor are more reactive than molecular species that adsorb at temperatures below 100°C [30, 33, 48]. It means that below 100°C, the thermal energy is not enough to produce the desorption reactions of the oxygen and, therefore, an

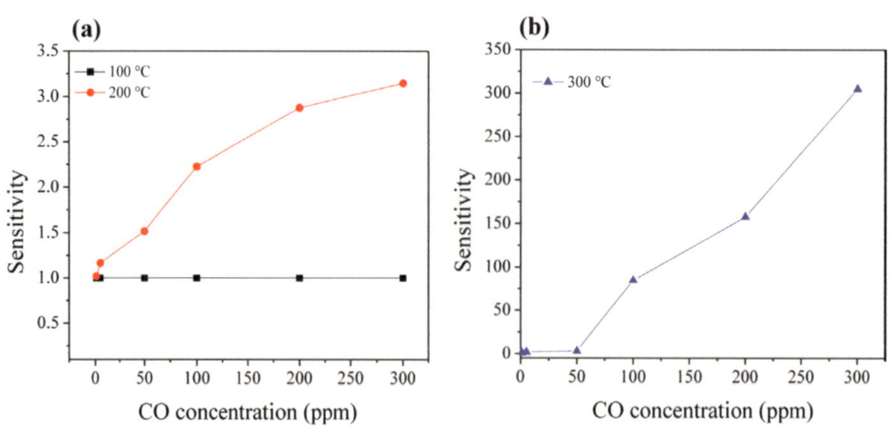

Figure 8. Sensitivity of the $ZnCo_2O_4$ sensor vs. concentration at different temperatures: (a) 100 and 200°C and (b) 300°C.

electrical response does not occur regardless of the gas concentration, as can see in **Figure 8a**. By contrast, above 100°C (in this case 200 and 300°C), the formation of ionic species at surface of the $ZnCo_2O_4$ occurs causing a chemical reaction with the gas and resulting in changes in the electrical resistance of the material (i.e., a high sensitivity is recorded) [48, 49]. Additionally, the conductivity mechanism of $ZnCo_2O_4$ sensor is strongly related to the crystallite size (*D*) and the space charge layer (*L*): if $D >> 2L$, the conductivity is limited by the Schottky barrier at the particle border; thus, gas detection does not depend on the size of the particle; if $D = 2L$, the conductivity and the gas sensing depend on the growing of necks formed by crystallites; and when $D < 2L$, the conductivity depends on the size of the crystallites [2]. In our case, the latter condition occurs while detecting the gases, since the average particle size is less than 100 nm; that is the reason why the conduction of the charge carriers (holes) takes place on the nanoparticles' surface [6, 50].

In comparison with previous works, our $ZnCo_2O_4$ sensor fabricated on faceted nanoparticles showed a superior sensitivity toward CO (a sensitivity of ~305 in 300 ppm of CO at 300°C) than those sensors based on $ZnCo_2O_4$ nanoparticles [18], hierarchical porous $ZnCo_2O_4$ nano/microspheres [19], $ZnCo_2O_4$ nanotubes [20], porous ZnO/$ZnCo_2O_4$ hollow spheres [21], and nanowires-assembled hierarchical $ZnCo_2O_4$ microstructure [25], with sensitivities of 1 (50 ppm at 350°C), 2 (100 ppm at 175°C), 1.69 (400 ppm at 300°C), 1.1 (100 ppm at 275°C), and 29 (10 ppm at 300°C), respectively.

2.7. Conclusions

$ZnCo_2O_4$-faceted nanoparticles (~75 nm) were obtained by the simple and inexpensive microwave-assisted colloidal method, using dodecylamine as surfactant. This synthetic method is allowed to obtain the $ZnCo_2O_4$ at a calcination temperature of 500°C. The sensing tests showed that $ZnCo_2O_4$ sensor is highly sensitive to concentrations of 1–300 ppm of carbon monoxide at working temperatures above 100°C. Specifically, a maximum sensitivity of 305 was obtained for a CO concentration of 300 ppm at a working temperature of 300°C. The CO sensing response of $ZnCo_2O_4$ is better than that reported in previous investigations. Therefore, $ZnCo_2O_4$ can be considered as a potential candidate for gas sensing applications.

Acknowledgements

The authors are grateful to PRODEP for financial support under the project F-PROMEP-39/Rev-04 SEP-23-005. The authors also thank PROFOCIE 2016 for financial support and CONACYT-México grants: the National Laboratory for Nanoscience and Nanotechnology Research (LINAN). The authors are grateful to G. J. Labrada-Delgado, B. A. Rivera-Escoto, K. Gomez, Miguel Ángel Luna-Arias and Sergio Oliva for their technical assistance.

Author details

Juan Pablo Morán-Lázaro[1*], Florentino López-Urías[2], Emilio Muñoz-Sandoval[2], Oscar Blanco-Alonso[3], Marciano Sanchez-Tizapa[4], Alejandra Carreon-Alvarez[4], Héctor Guillén-Bonilla[5], María de la Luz Olvera-Amador[6], Alex Guillén-Bonilla[1] and Verónica María Rodríguez-Betancourtt[7]

*Address all correspondence to: lazaro7mx27@gmail.com

1 Department of Computer Science and Engineering, CUValles, University of Guadalajara, Ameca, Jalisco, Mexico

2 Advanced Materials Department, IPICYT, San Luis Potosí, S.L.P., Mexico

3 Department of Physics, CUCEI, University of Guadalajara, Guadalajara, Jalisco, Mexico

4 Department of Natural and Exact Sciences, CUValles, University of Guadalajara, Ameca, Jalisco, Mexico

5 Department of Project Engineering, CUCEI, University of Guadalajara, Guadalajara, Jalisco, Mexico

6 Department of Electrical Engineering (SEES), CINVESTAV-IPN, Mexico City, DF, Mexico

7 Department of chemistry, CUCEI, University of Guadalajara, Guadalajara, Jalisco, Mexico

References

[1] Yamazoe N, Shimanoe K. New perspectives of gas sensor technology. Sens. Actuators B. 2009;**138**:100-107. DOI: 10.1016/j.snb.2009.01.023

[2] Xu C, Tamaki J, Miura N, Yamazoe N. Grain size effects on gas sensitivity of porous SnO_2-based elements. Sens. Actuators B. 1991;**3**:147-155. DOI: 10.1016/0925-4005(91)80207-Z

[3] Du J, Zhao R, Xie Y, Li J. Size-controlled synthesis of SnO_2 quantum dots and their gas-sensing performance. App. Surf. Sci. 2015;**346**:256-262. DOI: 10.1016/j.apsusc.2015.04.011

[4] Wan Q, Li QH, Chen YJ, Wang TH, He XL, Li JP, Lin CL. Fabrication and ethanol sensing characteristics of ZnO nanowire gas sensors. App. Phys. Lett. 2004;**84**:3654-3656. DOI: 10.1063/1.1738932

[5] Lin S, Wu J, Akbar SA. A selective room temperature formaldehyde gas sensor using TiO_2 nanotube arrays. Sens. Actuators B. 2011;**156**:505-509. DOI: 10.1016/j.snb.2011.02.046

[6] Kim HJ, Lee JH. Highly sensitive and selective gas sensors using p-type oxide semiconductors: overview. Sens. Actuators B. 2014;**192**:607-627. DOI: 10.1016/j.snb.2013.11.005

[7] Sharma Y, Sharma N, Subba-Rao GV, Chowdari BVR. Nanophase $ZnCo_2O_4$ as a high performance anode material for Li-ion batteries. Adv. Funct. Mater. 2007;**17**:2855-2861. DOI: 10.1002/adfm.200600997

[8] Du N, Xu Y, Zhang H, Yu J, Zhai C, Yang D. Porous $ZnCo_2O_4$ nanowires synthesis via sacrificial templates: high-performance anode materials of Li-ion batteries. Inorg. Chem. 2011;**50**:3320-3324. DOI: 10.1021/ic102129w

[9] Huang L, Waller GH, Ding Y, Chen D, Ding D, Xi P, Wang ZL, Liu M. Controllable interior structure of $ZnCo_2O_4$ microspheres for high-performance lithium-ion batteries. Nano Energy 2015;**11**:64-70. DOI: 10.1016/j.nanoen.2014.09.027

[10] Bai J, Li X, Liu G, Qian Y, Xiong S. Unusual formation of $ZnCo_2O_4$ 3D hierarchical twin microspheres as a high-rate and ultralong-life lithium-ion battery anode material. Adv. Funct. Mater. 2014;**24**:3012-3020. DOI: 10.1002/adfm.201303442

[11] Liu B, Zhang J, Wang X, Chen G, Chen D, Zhou C, Shen G, Qian Y, Xiong S. Hierarchical three-dimensional $ZnCo_2O_4$ nanowire arrays/carbon cloth anodes for a novel class of high-performance flexible lithium-ion batteries. Nano Lett. 2012;**12**:3005-3011. DOI: 10.1021/n1300794f

[12] Zhou G, Zhu J, Chen Y, Mei L, Duan X, Zhang G, Chen L, Wang T, Lu B. Simple method for the preparation of highly porous $ZnCo_2O_4$ nanotubes with enhanced electrochemical property for supercapacitor. Electrochim. Acta. 2014;**123**:450-455. DOI: 10.1016/j.electacta.2014.01.018

[13] Wu C, Cai J, Zhang Q, Zhou X, Zhu Y, Li L, Shen P, Zhang K. Direct growth of urchin-like $ZnCo_2O_4$ microspheres assembled from nanowires on nickel foam as high-performance electrodes for supercapacitors. Electrochim. Acta. 2015;**169**:202-209. DOI: 10.1016/j.electacta.2015.04.079

[14] Fu W, Li X, Zhao C, Liu Y, Zhang P, Zhou J, Pan X, Xie E. Facile hydrothermal synthesis of flower like $ZnCo_2O_4$ microspheres as binder-free electrodes for supercapacitors. Mater. Lett. 2015;**169**:202-209. DOI: j.matlet.2015.02.092

[15] Chang SK, Zainal Z, Tan KB, Yusof NA, Yusoff WMDW, Prabaharan SRS. Recent development in spinel cobaltites for supercapacitor application. Ceram. Int. 2015;**41**:1-14. DOI: 10.1016/j.ceramint.2014.07.101

[16] Gawande KB, Gawande SB, Thakare SR, Mate VR, Kadam SR, Kale BB, Kulkarni MV. Effect of zinc: cobalt composition in $ZnCo_2O_4$ spinels for highly selective liquefied petroleum gas sensing at low and high temperatures. RSC Adv. 2015;**5**:40429-40436. DOI: 10.1039/c5ra03960f

[17] Mariappan CR, Kumara R, Vijaya P. Functional properties of $ZnCo_2O_4$ nano-particles obtained by thermal decomposition of a solution of binary metal nitrates. RSC Adv. 2015;**5**:26843-26849. DOI: 10.1039/c5ra01937k

[18] Vijayanand S, Joy PA, Potdar HS, Patil D, Patil P. Nanostructured spinel $ZnCo_2O_4$ for the detection of LPG. Sens. Actuators B. 2011;**152**:121-129. DOI: 10.1016/j.snb.2010.09.001

[19] Liu T, Liu J, Liu Q, Song D, Zhang H, Zhang H, Wang J. Synthesis, characterization and enhanced gas sensing performance of porous $ZnCo_2O_4$ nano/microspheres. Nanoscale. 2015;**7**:19714-19721. DOI: 10.1039/c5nr05761b

[20] Zhang, GY, Guo B, Chen J. MCo_2O_4 (M = Ni, Cu, Zn) nanotubes: template synthesis and application in gas sensors. Sens. Actuators B. 2006;**114**:402-409. DOI: j.matlet. 2015.02.092

[21] Zhou X, Feng W, Wang C, Hu X, Li X, Sun P, Shimanoe K, Yamazoe N, Lu G. Porous $ZnO/ZnCo_2O_4$ hollow spheres: synthesis, characterization, and applications in gas sensing. J. Mater. Chem. A. 2014;**2**:17683-17690. DOI: 10.1039/c4ta04386c

[22] Niu X, Du W, Du W. Preparation and gas sensing properties of ZnM_2O_4 (M = Fe, Co, Cr). Sens. Actuators B. 2004;**99**:405-409. DOI: 10.1016/j.snb.2003.12.007

[23] Park HJ, Kim J, Choi NJ, Song H, Lee DS. Nonstoichiometric Co-rich $ZnCo_2O_4$ hollow nanospheres for high performance formaldehyde detection at ppb levels. ACS Appl. Mater. Interfaces. 2016;**8**:3233-3240. DOI: 10.1021/acsami.5b10862

[24] Qu F, Jiang H, Yang M. Designed formation through a metal organic framework route of $ZnO/ZnCo_2O_4$ hollow core–shell nanocages with enhanced gas sensing properties. Nanoscale. 2016;**8**:16349-16356. DOI: 10.1039/c6nr05187a

[25] Long H, Harley-Trochimczyk A, Cheng S, Hu H, Chi WS, Rao A, Carraro C, Shi T, Tang Z, Maboudian R. Nanowire-assembled hierarchical $ZnCo_2O_4$ microstructure integrated with a low-power microheater for highly sensitive formaldehyde detection. ACS Appl. Mater. Interfaces. 2016;**8**:31764-31771. DOI: 10.1021/acsami.6b11054

[26] Wei X, Chen D, Tang W. Preparation and characterization of the spinel oxide $ZnCo_2O_4$ obtained by sol–gel method. Mater. Chem. Phys. 2007;**103**:54-58. DOI: 10.1016/j.matchemphys.2007. 01.006

[27] Wang Y, Wang M, Chen G, Dong C, Wang Y, Fan LZ. Surfactant-mediated synthesis of $ZnCo_2O_4$ powders as a high-performance anode material for Li-ion batteries. Ionics. 2015;**21**:623-628. DOI: 10.1007/s11581-014-1221-1

[28] Morán-Lázaro JP, Blanco O, Rodríguez-Betancourtt VM, Reyes-Gómez J, Michel CR. Enhanced CO_2-sensing response of nanostructured cobalt aluminate synthesized using a microwave-assisted colloidal method. Sens. Actuators B. 2016;**226**:518-524. DOI: 10.1016/j. snb.2015.12.013

[29] Guillen-Bonilla H, Reyes-Gomez J, Guillen-Bonilla A, Pozas-Zepeda D, Guillen-Bonilla JT, Gildo-Ortiz L, Flores-Martinez M. Synthesis and characterization of $MgSb_2O_6$ trirutile-type in low presence concentrations of ethylenediamine. J. Chem. Chem. Eng. 2013;**7**: 395-401.

[30] Guillén-Bonilla A, Rodríguez-Betancourtt VM, Flores-Martínez M, Blanco-Alonso O, Reyes-Gómez J, Gildo-Ortiz L, Guillén-Bonilla H. Dynamic response of $CoSb_2O_6$ trirutile-type oxides in a CO_2 atmosphere at low-temperatures. Sensors. 2014;**14**:15802-15814. DOI: 10.3390/s140915802

[31] Blanco O, Morán-Lázaro JP, Rodríguez-Betancourtt VM, Reyes-Gómez J, Barrera A. Colloidal synthesis of $CoAl_2O_4$ nanoparticles using dodecylamine and their structural characterization. Superficies y Vacío. 2016;**29**:78-82.

[32] Guillen-Bonilla H, Rodríguez-Betancourtt VM, Guillén-Bonilla JT, Reyes-Gómez J, Gildo-Ortiz L, Flores-Martínez M, Olvera-Amador ML, Santoyo-Salazar J. CO and C_3H_8 sensitivity behavior of zinc antimonate prepared by a microwave-assisted solution method. J. Nanomater. 2015;**2015**:1-8. DOI: 10.1155/2015/979543

[33] Guillén-Bonilla H, Flores-Martínez M, Rodríguez-Betancourtt VM, Guillén-Bonilla A, Reyes-Gómez J, Gildo-Ortiz L, Olvera-Amador ML, Santoyo-Salazar J. A novel gas sensor based on $MgSb_2O_6$ nanorods to indicate variations in carbon monoxide and propane concentrations. Sensors. 2016;**16**:177-188. DOI: 0.3390/s16020177

[34] Guillén-Bonilla H, Gildo-Ortiz L, Olvera ML, Santoyo-Salazar J, Rodríguez-Betancourtt VM, Guillen-Bonilla A, Reyes-Gómez J. Sensitivity of mesoporous $CoSb_2O_6$ nanoparticles to gaseous CO and C_3H_8 at low temperatures. J. Nanomater. 2015;**2015**:1-9. DOI: 10.1155/2015/308465

[35] Mirzaei A, Neri G. Microwave-assisted synthesis of metal oxide nanostructures for gas sensing application: a review. Sens. Actuators B. 2016;**237**:749-775. DOI: 10.1016/j.snb.2016.06.114

[36] Hu SY, Lee YC, Chen BJ. Characterization of calcined $CuInS_2$ nanocrystals prepared by microwave-assisted synthesis. J. Alloys Compd. 2017;**690**:15-20. DOI: 10.1016/j.jallcom.2016.08.098

[37] Morán-Lázaro JP, López-Urías F, Muñoz-Sandoval E, Blanco-Alonso O, Sanchez-Tizapa M, Carreon-Alvarez A, Guillén-Bonilla H, Olvera-Amador ML, Guillén-Bonilla A, Rodríguez-Betancourtt. Synthesis, characterization, and sensor applications of spinel $ZnCo_2O_4$ nanoparticles. Sensors. 2016;**16**:2162. DOI: 10.3390/s16122162

[38] Ji Y, Zhao Z, Duan A, Jiang G, Liu J. Comparative study on the formation and reduction of bulk and Al_2O_3-supported cobalt oxides by H_2-TPR technique. J. Phys. Chem. C. 2009;**113**:7186-7199. DOI: 10.1021/jp8107057

[39] LaMer VK, Dinegar RH. Theory, production and mechanism of formation of monodispersed hydrosols. J. Am. Chem. Soc. 1950;**72**:4847-4854. DOI: 10.1021/ja01167a001

[40] Vekilov PG. What determines the rate of growth of crystals from solution? Cryst Growth Des. 2007;**7**:2796-2810. DOI: 10.1021/cg070427i

[41] Samanta K, Bhattacharya P, Katiyar RS. Raman scattering studies in dilute magnetic semiconductor $Zn_{1-x}Co_xO$. Phys. Rev. B. 2006;**73**:245213-245215. DOI: 10.1103/PhysRevB.73.245213

[42] Shirai H, Morioka Y, Nakagawa I. Infrared and Raman spectra and lattice vibrations of some oxide spinels. J. Phys. Soc. Jpn. 1982;**51**:592-597. DOI: 10.1143/JPSJ.51.592

[43] Chang SC. Oxygen chemisorption on tin oxide: correlation between electrical conductivity and EPR measurements. J. Vac. Sci. Technol. 1979;**17**:366-369. DOI: 10.1116/1.570389

[44] Gildo-Ortiz L, Guillén-Bonilla H, Santoyo-Salazar J, Olvera ML, Karthik TVK, Campos-González E, Reyes-Gómez J. Low-temperature synthesis and gas sensitivity of perovskite-type $LaCoO_3$ nanoparticles. J. Nanomater. 2014;**2014**:1-8. DOI: 10.1155/2014/164380

[45] Yamazoe N. New approaches for improving semiconductor gas sensors. Sens. Actuators B. 1991;**5**:7-19. DOI: 10.1016/0925-4005(91)80213-4

[46] Moseley PT. Materials selection for semiconductor gas sensors. Sens. Actuators B. 1992;**6**:149-156. DOI: 10.1016/0925-4005(92)80047-2

[47] Bochenkov VE, Sergeev GB. Preparation and chemiresistive properties of nanostructured materials. Adv. Colloid Interface Sci. 2005;**116**:245-254. DOI: 10.1016/j.cis.2005.05.004

[48] Wang C, Yin L, Zhang L, Xiang D, Gao R. Metal oxide gas sensors: sensitivity and influencing factors. Sensors. 2010;**10**:2088-2106. DOI: 10.3390/s100302088

[49] Yamazoe, N. Toward innovations of gas sensor technology. Sens. Actuators B. 2005;**108**:2-14. DOI: 10.1016/j.snb.2004.12.075

[50] Tan OK, Cao W, Hu Y, Zhu W. Nano-structured oxide semiconductor materials for gas-sensing applications. Ceram. Int. 2004;**30**:1127-1133. DOI: 10.1016/j.ceramint.2003.12.015

Chapter 9

Aluminum- and Iron-Doped Zinc Oxide Nanorod Arrays for Humidity Sensor Applications

Ahmad Syakirin Ismail, Mohamad Hafiz Mamat and Mohamad Rusop Mahmood

Additional information is available at the end of the chapter

http://dx.doi.org/10.5772/67661

Abstract

Metal-doped zinc oxide (ZnO) nanorod arrays have attracted much attention due to improvement in their electrical, structural, and optical properties upon doping. In this chapter, we discuss the effects of aluminum (Al)- and iron (Fe)-doping on ZnO nanorod arrays properties particularly for humidity sensor applications. Compared to Fe, Al shows more promising characteristics as doping element for ZnO nanorod arrays. The Al-doped ZnO nanorod arrays showed dense arrays, small nanorods diameter, and high porous surface. The I-V characteristics showed that Al-doped sample possesses higher conductivity. From the humidity sensing performance of the samples, Al-doped ZnO nanorod arrays possess the superior sensitivity, more than two times higher than that of the undoped ZnO nanorod arrays sample, demonstrating great potential of Al-doped ZnO nanorod arrays in humidity sensor applications.

Keywords: ZnO, Al-doped, Fe-doped, nanorod arrays, humidity sensors

1. Introduction

Zinc oxide (ZnO) has been the subject of extensive investigations for many decades. A stable hexagonal wurtzite structure of ZnO consists of tetrahedrally coordinated four –O or four –Zn atoms, having lattice constant of a = 3.25 Å and c = 5.2 Å with ratio c/a = ~1.60 close to ideal hexagonal cell (1.633) [1]. **Figure 1** shows the ball and stick illustration of the hexagonal wurtzite structure of ZnO [1]. The ZnO structure does not have the center of symmetry.

ZnO is abundantly available in nature as a mineral zincite, although most of the commercial ZnO samples are prepared through synthetic approaches. Owing to its interesting properties,

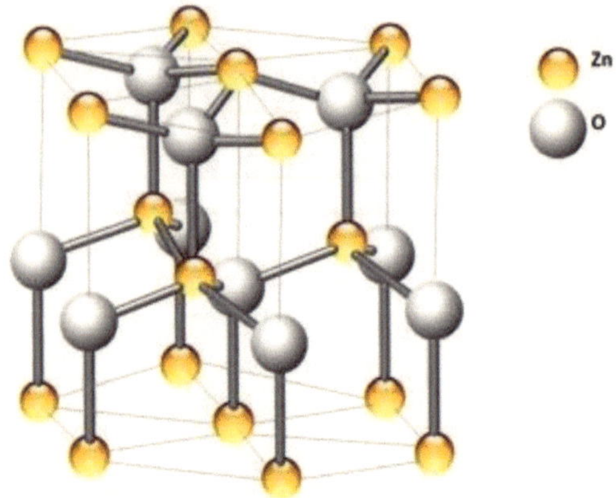

Figure 1. The illustration of the hexagonal wurtzite structure of ZnO.

such as a wide energy band gap of 3.37 eV, large exciton binding energy of 60 meV, good chemical and thermal stabilities, non-toxicity, and high transparency [2–5], ZnO has been applied in various applications such as textiles [6], medicines [7, 8], optoelectronic devices [9], solar cells [10, 11], and sensors [12, 13]. Other promising aspect of ZnO includes its ease of fabrication leading to generation of various kinds of nanostructures such as nanorods [14], nanowires [15], nanoflowers [16], and nanospheres [17]. Among them, one-dimensional (1-D) nanostructures (i.e., nanorods and nanowires) are more favorable in humidity sensor applications due to a higher surface-to-volume ratio, providing direct charge transport along the ZnO arrays and reducing the electron-hole pair recombination possibilities [18, 19]. These characteristics are crucial in fabricating high-performance humidity sensors. The performance of humidity sensor can be enhanced by introducing impurities to ZnO crystal through the doping process. As reported in previous studies, the doping process is capable of altering electrical conductivity, photocatalytic activity, and magnetic properties of ZnO [20–23]. In this chapter, we discuss recent developments in metal-doped ZnO, particularly in relation to the advantages of nanostructure-based humidity sensors. The preparation of Al- and Fe-doped ZnO nanorod array-based humidity sensors using the sol-gel immersion method is also included here followed by characterization of the structural, optical, and electrical properties of the fabricated humidity sensors.

2. Metal-doped ZnO

Doping is a process of introducing extrinsic elements into intrinsic structure for the purpose of improving and altering their basic properties such as structural, optical, and electrical. This process is crucial for various materials especially semiconductors since intrinsic semiconductor is known to possess deficiencies such as low carrier concentrations and lagging change in resistance values for humidity sensing applications [24, 25]. From the literature, ZnO was reported to be doped with numerous kinds of elements [26–29]. For instance, Kim et al. reported on the

http://dx.doi.org/10.5772/67661

fabrication of yttrium (Y)-doped ZnO nanorod arrays quantum dot (QD) synthesized solar cell using the chemical bath deposition method [30]. Due to Y-doping, the nanorod diameters were significantly reduced. In addition, the absorption coefficients of the films also improved after doping due to higher QD deposition. The conductivity of the film was improved when the concentration of doping was 50 mM. They also observed that the undoped ZnO nanorod arrays possess poor solar cell characteristics than that of the 50 mM Y-doped ZnO nanorod arrays sample. The substitution of the Y atom in ZnO lattice was expected to affect the transport properties of electrons by blocking the recombination process and increased the carrier concentration.

In another work, Kim et al. studied the fabrication of hydrophobic Al-doped ZnO nanorod arrays solar cell using the hydrothermal method [31]. In this study, they varied the concentration of Al dopant and controlled the growth period. The average diameters were significantly reduced when doped with Al. In addition, they reported that the undoped sample has higher hysteresis than that of Al-doped samples and the Al-doped ZnO nanorod arrays have a high self-cleaning performance. Meshki et al. fabricated Fe-doped ZnO nanorod arrays for two kinds of drug detection, namely, sulfamethoxazole and sulfamethizole [32]. When the nanorods were doped with Fe, the crystallite size of the ZnO nanorod arrays film was slightly decreased. The Fe-doped ZnO nanorods acquired a larger surface area than undoped ZnO nanorods, which likely led to the enhancement of electron transfer to the electrode surface.

In other research, Anbia and Fard fabricated a cerium (Ce)-doped ZnO nanoporous thin film humidity sensor using the screen printed method [33], in which they reported changes in humidity-sensing performance with doping and different sintering temperature. From their research, the Ce-doped ZnO thin film acquired smaller diameter of nanoparticles with a high porous surface. They also found that the Ce-doped ZnO film has better sensitivity to humidity than that of the undoped ZnO film. Peng et al. fabricated manganese (Mn)-doped ZnO nanopowders for humidity sensing application [34]. Based on their study, the undoped ZnO nanopowders possess larger nanoparticles and have low porosity. Upon Mn doping, the nanoparticle size is reduced with higher porous surface. As the concentration of Mn increased, the sensitivity of the ZnO nanopowders to humidity also increased, clearly higher than the undoped ZnO powder. They expected such behavior to occur due to higher H^+ ion density on the ZnO nanopowders surface and also higher concentration of defects with associated with oxygen vacancies.

From the above summary of the reported results, it is evident that doping is essential and very important for improving the useful properties of pristine ZnO structure. The improvements such as reduced crystallite size (increase surface area), high-quality surface structure, better photoluminescence properties, lower surface resistance, and high concentration of free carrier enhance the efficiency of the ZnO-based devices.

3. ZnO nanostructure-based humidity sensors

Humidity sensors have an important function in various situations, such as in industrial processing, agriculture, and environmental control since an uncontrolled amount of humidity can be quite damaging to certain materials and processes [35]. Humidity sensors can be based on different sensing techniques, such as resistive, capacitive, optics, field effect transistor (FET), quartz crystal microbalance (QCM), and surface acoustic wave (SAW) [36–40]. Resistive-type humidity sensors are often preferred since they are easy to fabricate, have low cost, and have reliable response [41]. ZnO nanostructures have been widely studied for humidity sensing applications since it is highly sensitive to humidity. For example, Hendi et al. investigated the effect of different concentrations of an Sn-doped ZnO-based QCM humidity sensor [42]. They found that the crystallite sizes decrease when doped with Sn and the performance of the QCM humidity sensor improved when doped with Sn. Zhao et al. [43] have fabricated gallium (Ga)-doped ZnO nanowires piezo-humidity sensor using hydrothermal method. At low concentration of doping, they reported that the nanowires possess a small average diameter and doping with Ga helped to increase the oxygen vacancies and generate more free carriers, which in turn changed the piezoelectric screening effect and improved the water molecules adsorption. These results showed that Ga-doped ZnO nanowires yielded higher response to humidity compared to undoped ZnO nanowires.

The nano-scale structure in ZnO is very important in humidity detection due to larger surface-to-volume ratio and chemically reactive surface. According to Hsu et al. [44], the nanostructured surface of ZnO consists of high concentrations of oxygen vacancies, which provides highly active sites for water molecules adsorption. In their study, they fabricated ZnO dandelion-like nanostructures using a two-step thermal oxidation method, reporting increment in oxygen vacancies as the deposition temperature increased, which in-turn led to enhancement of humidity sensitivity.

High surface area is one of key factors that determines the high performance of a humidity sensor. The basic mechanism of a humidity sensor involves water molecules getting attached to the ZnO surface following the Grotthuss chain reaction [45]:

$$2H_2O \rightarrow H_3O^+ + OH^- \quad (1)$$

$$H_3O^+ \rightarrow H_2O + H^+ \quad (2)$$

The source of charge carriers is from the protonic transfer (H^+) among hydronium ions (H_3O^+) which is known by the term proton hopping. A higher surface area offers more surface reactions between the sensing element and water molecules leading to a greater number of carriers. This behavior was observed by Hong et al. when preparing ZnO nanorod arrays on top of the gallium (Ga)-doped ZnO seed layer film through the hydrothermal method [46]. They suggested that the increment concentrations of Ga in the ZnO seed layer film increased the thickness but reduced the nanorod diameter, resulting in an increased surface area of the nanorod arrays film and hence higher sensitivity for humidity detection.

4. Synthesis of zinc oxide nanorod arrays

For the synthesis of ZnO nanorod array films, a number of approaches have been reported. For instance, Ye et al. synthesized the sodium (Na)-doped ZnO nanorods on silicon (Si) substrate using the chemical vapor deposition method [14] and studied the effects of Na concentrations on the ZnO nanorod properties. The growth of ZnO nanorods was done at 1000°C and 1800 Pa of temperature and pressure, respectively. Based on their report, the synthesized nanorods have a small diameter but they are not well aligned. In the report by Li et al. [47], the nanorod film was prepared with different dopants (Ga and Indium (In)) using the hydrothermal method on top of the ZnO seed layer-coated glass substrate. The purpose of their study was to investigate the influence of hydrogen annealing on the structural and luminescence properties of the ZnO nanorods. Through the preparation process, the nanorods were grown on the substrate with poor orientation and not aligned. According to them, the roughness of nanorods surface was increased due to thermal decomposition of the unstable surface status of polar and nonpolar faces of the ZnO structure. In the work by Son et al [48], the chemical bath deposition method was used to study the effect of seed layer thickness which was sputtered on Si substrate to the nanorod growth. Their results showed that the nanorods grew in a nonuniform manner in that the thickness and diameters of the nanorods were not uniform. However, higher thickness of the seed layer film produced nanorod arrays with better uniformity and alignment on the substrate.

From a review of the above-cited examples, it is evident that an ideal preparation of high-quality nanorod arrays does not necessarily depend on high-applied temperature and pressure. Therefore, a simpler, lower cost, and low-energy consumption method would be preferable. In our study, the synthesis of Al- and Fe-doped ZnO nanorod arrays was accomplished using a simple sol-gel immersion method. The nanorod arrays were grown on top of glass substrate with Al-doped ZnO as a seeded catalyst layer. For the preparation of the ZnO seeded layer catalyst, 0.4 M zinc acetate dehydrate ($Zn(CH_3COO)_2{\cdot}2H_2O$; 99.5% purity; Aldrich) was used as a precursor, 0.4 M mono-ethanolamine (MEA, $H_2NCH_2CH_2OH$; 99.5% purity; Aldrich) as a stabilizer, 0.004 M aluminum nitrate nonahydrate ($Al(NO_3)_3.9H_2O$; 98% purity; Analar) was used as a dopant source and 2-methoxyethanol as the solvent. The materials were mixed and then stirred on a hot plate stirrer at 80°C for 3 h and then aged for 24 h at room temperature to obtain a homogeneous solution. The prepared solution was spin coated at 3000 rpm for 1 min. The samples with coated film were then heated at 150°C for 10 min to dry and annealed at 500°C for 1 h.

The major interest of this study is the effect of Al and Fe doping on the intrinsic properties of ZnO. For the preparation of an undoped, Al-, and Fe-doped ZnO nanorod array film, 0.1 M zinc nitrate hexahydrate ($Zn(NO_3)_2{\cdot}6H_2O$; 98.5% purity; Schmidt) was used as a precursor, 0.1 M hexamethylenetetramine (HMT, $C_6H_{12}N_4$; 99% purity; Aldrich) as a stabilizer, and 0.001 M of aluminum nitrate dehydrate ($Al(NO_3)_3{\cdot}9H_2O$; 98% purity; Analar) and 0.001 M iron (III) nitrate nonahydrate ($Fe(NO_3)_3{\cdot}9H_2O$; 98% purity; Merck) as dopant sources. The reagents were mixed and dissolved in 500 ml of deionized water and then sonicated at 50°C for 30 min before being stirred on a magnetic hot plate stirrer at room temperature for 3 h to age the solutions. After the ageing process, the solutions were poured in a Schott bottle placed with seed layer-coated glass substrates. The seed layer-coated glass substrates were immersed in a water bath immersion tank at 95°C for 2 h. Then, the sample was cleaned and heated at 150°C for 10 min before being annealed at 500°C for 1 h.

The configuration of doped and undoped ZnO nanorod array-based humidity sensors was completed by depositing gold (Au) metal contact using sputter coater on top of the samples as the electrode. The structural properties of the samples were characterized using field emission scanning electron microscopy (FESEM; JEOL JSM-7600F) and X-ray diffraction measurement (XRD; PANalytical X'Pert PRO). The optical properties of the samples were tested using a ultraviolet-visible-near-infrared (UV-vis-NIR; Varian Cary 5000) spectrophotometer. The electrical properties of the samples were characterized using two-point probe current-voltage (I-V; Advantest R6243) measurement. The performance of the humidity sensor of the fabricated sensors was analyzed using a humidity sensor measurement system (ESPEC-SH261). The schematics of preparation of doped and undoped ZnO nanorod array-based humidity sensors are shown in **Figure 2**.

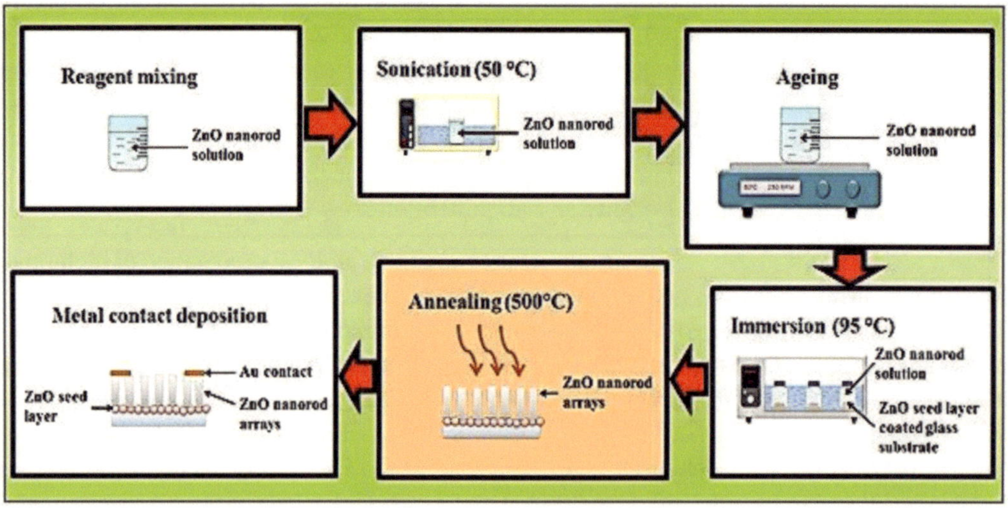

Figure 2. Schematics of the preparation of doped and undoped ZnO nanorod array-based humidity sensors.

5. Effects of aluminum and iron doping

The surface morphology, thickness, and the elemental analysis of the ZnO nanorod arrays were investigated using field emission electron scanning microscopy (FESEM) and energy-dispersive X-ray spectroscopy (EDS) as shown in **Figure 3**. **Figure 3(a–c)** shows the surface morphology of undoped and doped ZnO nanorod arrays taken at 30,000× magnification. It is observed that all samples produced have a hexagonal nanorod structure. The undoped sample has the largest average diameter of nanorod arrays of about 100 nm while Al- and Fe-doped ZnO-produced nanorod arrays have average diameters of 65 and 90 nm, respectively. Besides, the Al-doped ZnO nanorod array film is observed to possess higher porosity compared to undoped and Fe-doped samples. The reduction of nanorods size due to Al and Fe doping may be due to the smaller ionic radius of Al^{3+} (0.54 Å) and Fe^{3+} (0.64 Å) ions substituting the Zn^{2+} (0.74 Å) ion sites [49, 50]. The cross-sectional images of Al- and Fe-doped ZnO nanorod arrays are shown in **Figure 3(d–f)**. The thickness of undoped, Al-, and Fe-doped ZnO nanorod arrays films was estimated to be 1.1, 0.67, and 0.75 μm, respectively. A slight decrement in thickness of doped samples may be due to difference in the ionic radius as mentioned earlier. From

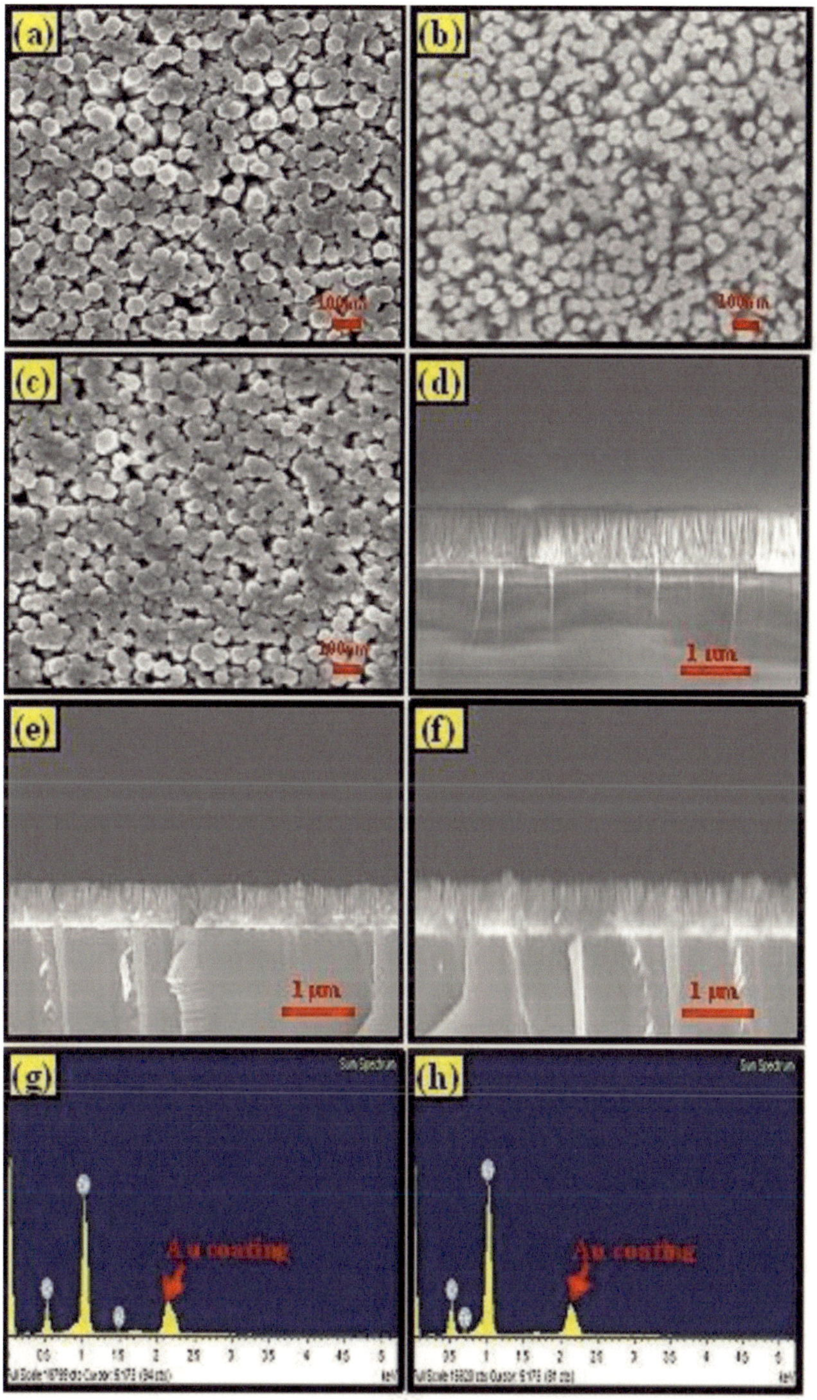

Figure 3. Surface morphology of (a) undoped, (b) Al-, and (c) Fe-doped ZnO nanorod array films that are prepared using sonicated sol-gel immersion. Cross-sectional images of (d) undoped, (e) Al-, and (f) Fe-doped ZnO nanorod array films. EDX spectrum of (g) Al- and (h) Fe-doped ZnO nanorod array films.

the EDS spectrum in **Figure 3(g** and **h)**, it is evident that the doping elements of Al and Fe appeared in each sample. The atomic ratio of Zn, Al, and O is expected to be 54.93:0.71:44.86, respectively, while the atomic ratio of Zn, Fe, and O is 53.60:0.21:46.18, respectively.

The XRD patterns of the undoped, Al-, and Fe-doped ZnO nanorod arrays for the 2θ range of 20–60° in **Figure 4** show the four main peaks with Miller indices of (100), (002), (101), and (102). The dominant (002) peak indicates that the growth is along *c*-axis orientation. In addition, it is also noticed that the diffraction peak at (002) orientation for doped samples is slightly decreased, which indicated that the doping elements are successfully substituting the ZnO structure. The degradation in peak intensities for doped samples was expected due to the difference in ionic radii of Al^{3+} and Fe^{3+} than that of Zn^{2+} ions. The position of the (002) peak for the undoped, Al-, and Fe-doped ZnO nanorod arrays occurs at 2θ = 34.38, 34.49, and 34.44°, respectively, which can be used to determine the lattice constant *c* using the following equation valid for hexagonal structure [51, 52]:

$$c = \frac{\lambda}{\sin\theta} \tag{3}$$

Here, λ = 1.54 Å is the X-ray wavelength of CuKα radiation. For the undoped, Al-, and Fe-doped ZnO nanorod arrays, *c* = 5.2108, 5.1947, and 5.2020 Å, respectively, was determined. The crystallite sizes, *D* of the undoped, Al-, and Fe-doped ZnO nanorod arrays, were calculated using the Scherrer formula [53]:

$$D = \frac{K\lambda}{\beta\cos\theta} \tag{4}$$

Here, *K* is a constant (0.94), and β is the full-width at half-maximum (FWHM) in radians. The (002) peak was used for the calculation of *D* for which the values of β of the undoped, Al-, and Fe-doped ZnO nanorod arrays are 0.2831, 0.3045, and 0.3037°, respectively. The *D* values of the undoped, Al-, and Fe-doped ZnO nanorod arrays were estimated to be 30.7, 28.5, and 28.6 nm, respectively. Thus, the values of *c* and *D* of the undoped ZnO nanorod arrays are lowered when doped with Al and Fe. According to Khuili et al. [54], lattice parameters *a* and *c* of Al-doped ZnO are smaller than those for undoped ZnO due to substitution of smaller Al^{3+} ions

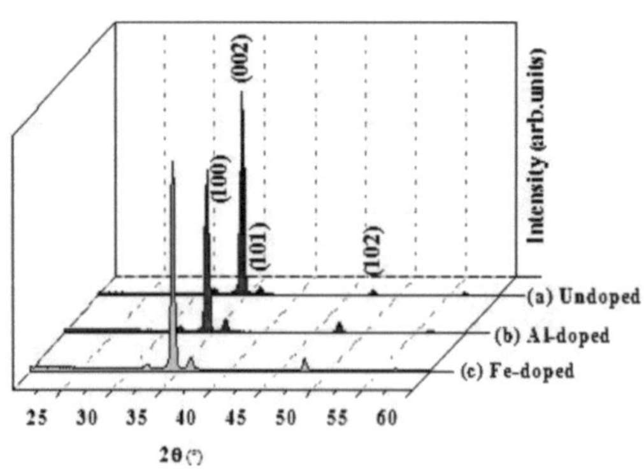

Figure 4. The XRD patterns of the undoped, Al-, and Fe-doped ZnO nanorod arrays.

http://dx.doi.org/10.5772/67661

into ZnO lattice, which is also confirmed by the simulated supercell structure. Yue et al. noted that the shrinkage of lattice constant and crystallite size when doped with Fe was due to the smaller ionic radius of Fe^{3+} compared to that of Zn^{2+} ions [50]. Bai et al. also found that the diffraction peak positions of ZnO have shifted to higher angles when doped with Fe, which might lead to reduction of lattice constant and crystallite size [55].

The optical properties of undoped and doped ZnO nanorod arrays were determined using UV-Vis-NIR spectrophotometer measurements between 350 and 800 nm at room temperature. **Figure 5** shows the transmittance properties of undoped, Al-, and Fe-doped ZnO nanorod arrays. The transmission decreases significantly at approximately 380 nm, which is attributed to the intrinsic ZnO band gap because of the direct transition of electrons between the edges of the valence band to the conduction band. From the spectra, average transmittances of undoped, Al-, and Fe-doped ZnO nanorod arrays were estimated to be 78.59, 76.96, and 75.10%, respectively. All doped samples exhibit a decrement in transmittance compared to the undoped sample. Previous studies reported that the transmittance decreased with Al and Fe doping [56, 57]. Such behavior occurred due to reduction of crystalline properties, which can be observed from the XRD data, and also due to enhancement of optical scattering by grain boundaries [57].

Figure 6 shows the I-V characteristics of undoped, Al-, and Fe-doped ZnO nanorod arrays. The I-V curves indicate that the nanorod arrays exhibit ohmic behavior, and the current increases with increasing voltage supplied for the nanorod arrays. The conductivity of the film, σ, was determined using the following equation [51]:

$$\sigma = \frac{1}{\rho}, \tag{5}$$

where ρ is the resistivity that can be expressed as

$$\rho = \left(\frac{V}{I}\right)\frac{wt}{l}, \tag{6}$$

where V is the supplied voltage, I is the measured current, t is the film's thickness, w is the electrode width, and l is the length between the electrodes. The active area of the thin films is

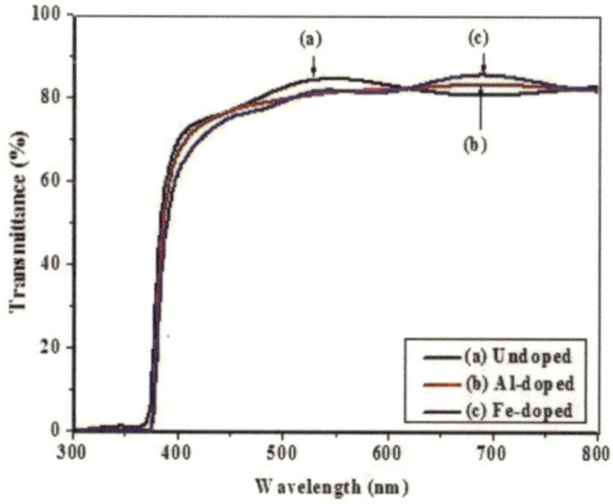

Figure 5. Transmittance properties of the undoped, Al-, and Fe-doped ZnO nanorod arrays.

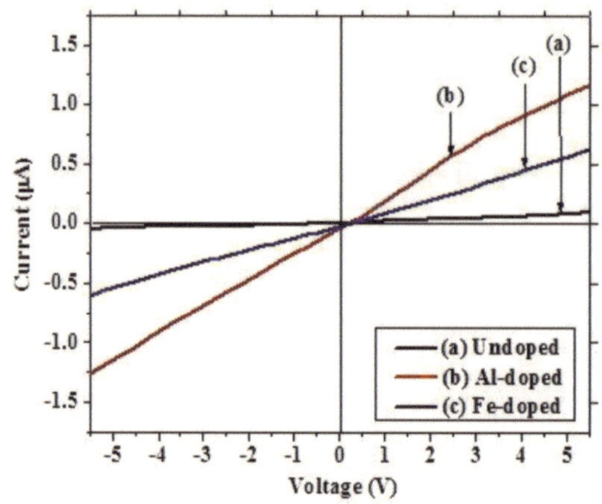

Figure 6. I-V plot for the undoped, Al-, and Fe-doped ZnO nanorod arrays.

3×10^{-6} m^2. The conductivity of undoped, Al-, and Fe-doped ZnO nanorod arrays is 0.08, 0.80, and 0.64 S cm^{-1}, respectively. It is observed that the conductivity of the nanorod array film significantly increased after doping. When ZnO was doped with Al^{3+} ions, the generation of free carriers is as follows [58]:

$$Al_2O_3 \rightarrow Al_{Zn\bullet} + 2O_o + \frac{1}{2}O_2 + 2e' \qquad (7)$$

where $Al_{Zn}^{\bullet}$ represents one positive charge of Al that stayed in zinc lattice and act as donor. O_o is the oxygen ion in the inherent lattice. On the other hand, when ZnO was doped with Fe^{3+} ions, the generation of free carriers is as follows [55]:

$$Fe_2O_3 \rightarrow 2Fe_{Zn}^{\bullet} + 3O_o^x + 2e' \qquad (8)$$

where $Fe_{Zn}^{\bullet}$ represents one positive charge of Fe that stayed in zinc lattice and act as donor. O_o is the oxygen ion in the inherent lattice while "x" represents the neutrality of O_o. Substitutions of two Fe atoms in the ZnO structure have induced two free electrons. This theory supports the enhancement of conductivity from Al- and Fe-doped samples compared to undoped ZnO nanorod arrays.

Humidity sensor responses of the prepared sensors in ambient with various RH (relative humidity) levels are shown in **Figure 7**. The measurements were conducted with 5 V DC-biased source at room temperature. During the adsorption process (40–90% RH), current signal increased steadily with increasing RH. On the other hand, during the desorption process (90–40% RH), current signal rapidly dropped until the signal recovers back to its initial current value. Response and recovery times are among the most imperative elements for evaluating the performance of humidity sensors. It is noted that the time taken by a sensor to achieve 90% of the total current change is defined as the response time in the case of adsorption or the recovery time in the case of desorption. The response/recovery times of the undoped, Al-doped, and Fe-doped ZnO nanorod arrays were measured to be 252/360, 270/90, and 243/108 s, respectively. Response and recovery behavior of sensor

http://dx.doi.org/10.5772/67661

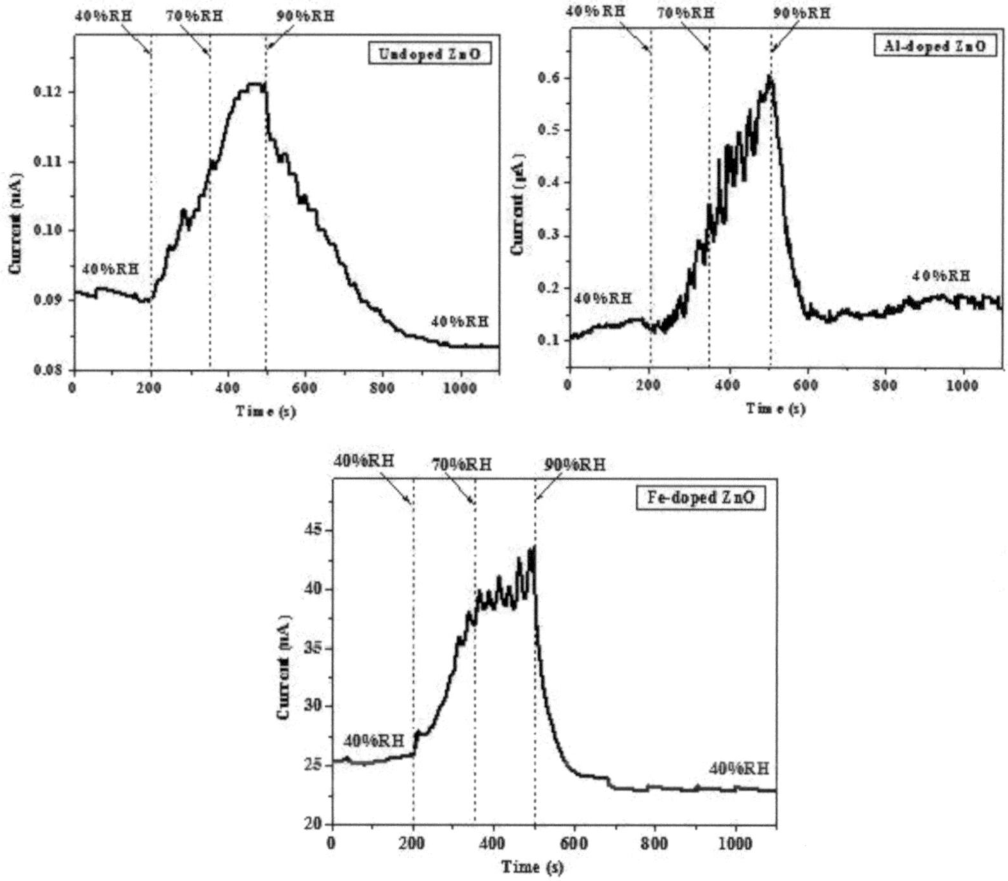

Figure 7. Humidity sensor response of (a) undoped, (b) Al- doped, and (c) Fe-doped ZnO nanorod array-based humidity sensors.

is dependent on large surface-to-volume ratio, where it helps to increase the adsorption of water molecules on the sensing element surface. Besides, with the availability of 1 D nanostructures (nanorod), it provides a large length-to-diameter ratio which can accelerate water molecules transfer to and from the interaction region as well as enhancing electron transfer along them [59].

Another important element of the humidity sensor is the sensitivity of the sensor. The sensitivity of undoped and doped ZnO nanorod array-based humidity sensors was estimated by using resistance data obtained from the response curve. Sensitivity was calculated by using the following relation [60, 61]:

$$S = \frac{R_a}{R_{rh}} \tag{9}$$

where S is sensitivity, R_a is the resistance of the sensor under exposure to the initial humidity level, and R_{rh} is the resistance of the sensor at the maximum humidity level. The values of resistance were obtained using Ohm's law on the basis of a fixed bias voltage of 5 V, shown as follows:

$$V = IR \tag{10}$$

where V is the bias voltage, I is the measured current, and R is the resistance. The values of R_a/R_{rh} for undoped, Al-doped, and Fe-doped ZnO nanorod array-based humidity sensors were calculated to be $5.56 \times 10^9/4.17 \times 10^9$, $2.66 \times 10^7/8.33 \times 10^6$, and $2.08 \times 10^8\,\Omega/1.19 \times 10^8\,\Omega$, respectively. From R_a and R_{rh} values, the sensitivities of undoped, Al-doped, and Fe-doped ZnO nanorod array-based humidity sensors were calculated to be 1.33, 3.19, and 1.75, respectively. All doped samples show an improved performance in humidity sensing compared to the undoped sample. Zhu et al. also noticed the improvement of humidity sensing capabilities when they fabricated the piezoelectric-based ZnO nanowires humidity sensor [62]. In other report, Hendi et al. reported the response/recovery time of the ZnO-based QCM humidity sensor was improved when they doped ZnO with Sn [42]: this was attributed to easy diffusion of water molecules between ZnO nanopowders.

Zhu et al. reported that the doping process can increase the concentration of oxygen vacancies when the doping elements substituted the ZnO lattice which improved the performance of humidity sensor [62]. In addition, the substitution of Al^{3+} and Fe^{3+} ions has increased the charge density and offering more surface reaction with water molecules. This statement is supported by increase in conductivity and E_g values as discussed earlier. Furthermore, the superior performance of Al-doped ZnO nanorod arrays is likely related to the higher porosity of its surface as observed from FESEM image.

6. Concluding remarks

In this chapter, advantages of doping of zinc oxide (ZnO) nanostructures have been reviewed along with changes in their properties upon doping, particularly in relation to humidity sensing applications. Compared to Fe-doping, Al-doping shows more useful changes in properties for ZnO nanorod arrays for sensor applications. From the FESEM images, the Al-doped ZnO nanorods showed dense arrays with an average diameter of 65 nm, and higher porosity of the surface as compared to the Fe-doped sample which has an average diameter of 90 nm. No significant changes to the optical properties of doped samples were observed while the I-V characteristic of those doped samples possess a significant increment in conductivity compared to undoped ZnO nanorod arrays. Regarding the humidity-sensing performance of the samples, Al-doped ZnO nanorod arrays showed superior sensitivity, more than two times higher than that of the undoped ZnO nanorod array sample. These results show that Al-doped ZnO nanorod arrays has a promising future in humidity sensor applications.

Acknowledgements

This work was supported by the Fundamental Research Grant Scheme 600-RMI/FRGS 5/3 (57/2015) from the Ministry of Education Malaysia. The authors also would like to thank the Faculty of Electrical Engineering, NANO-ElecTronic Centre (NET), NANO-SciTech Centre

http://dx.doi.org/10.5772/67661

(NST), Research Management Centre (RMC) of UiTM and the Ministry of Higher Education of Malaysia for their funding, laboratory equipment, and support of this research.

Author details

Ahmad Syakirin Ismail[1*], Mohamad Hafiz Mamat[1,2] and Mohamad Rusop Mahmood[1,2]

*Address all correspondence to: kyrin_samaxi@yahoo.com

1 NANO-ElecTronic Centre (NET), Faculty of Electrical Engineering, Technology University Mara (UiTM), Shah Alam, Selangor, Malaysia

2 NANO-SciTech Centre (NET), Institute of Science (IOS), Technology University Mara (UiTM), Shah Alam, Selangor, Malaysia

References

[1] Özgür Ü, Alivov YI, Liu C, Teke A, Reshchikov MA, Doğan S, et al. A comprehensive review of ZnO materials and devices. Journal of Applied Physics. 2005;**98**:041301

[2] Al-Kuhaili MF, Durrani SMA, El-Said AS, Heller R. Enhancement of the refractive index of sputtered zinc oxide thin films through doping with Fe_2O_3. Journal of Alloys and Compounds. 2017;**690**:453-460

[3] Li C-C, Jhang J-H, Tsai H-Y, Huang Y-P. Water-soluble polyethylenimine as an efficient dispersant for gallium zinc oxide nanopowder in organic-based suspensions. Powder Technology. 2017;**305**:226-231

[4] Ahmad H, Lee CSJ, Ismail MA, Ali ZA, Reduan SA, Ruslan NE, et al. Zinc oxide (ZnO) nanoparticles as saturable absorber in passively Q-switched fiber laser. Optics Communications. 2016;**381**:72-76

[5] Nazarkovsky MA, Bogatyrov VM, Czech B, Galaburda MV, Wójcik G, Kolomys OF, et al. Synthesis and properties of zinc oxide photocatalyst by high-temperature processing of resorcinol-formaldehyde/zinc acetate mixture. Journal of Photochemistry and Photobiology A: Chemistry. 2017;**334**:36-46

[6] Pulit-Prociak J, Chwastowski J, Kucharski A, Banach M. Functionalization of textiles with silver and zinc oxide nanoparticles. Applied Surface Science. 2016;**385**:543-553

[7] Chen J, Zhang X, Cai H, Chen Z, Wang T, Jia L, et al. Osteogenic activity and antibacterial effect of zinc oxide/carboxylated graphene oxide nanocomposites: Preparation and in vitro evaluation. Colloids and Surfaces B: Biointerfaces. 2016;**147**:397-407

[8] Alswat AA, Ahmad MB, Saleh TA, Hussein MZB, Ibrahim NA. Effect of zinc oxide amounts on the properties and antibacterial activities of zeolite/zinc oxide nanocomposite. Materials Science and Engineering: C. 2016;**68**:505-511

[9] Gupta N, Grover R, Mehta DS, Saxena K. A simple technique for the fabrication of zinc oxide-PEDOT: PSS nanocomposite thin film for OLED application. Synthetic Metals. 2016;**221**:261-267

[10] Jia X, Wu N, Wei J, Zhang L, Luo Q, Bao Z, et al. A low-cost and low-temperature processable zinc oxide-polyethylenimine (ZnO:PEI) nano-composite as cathode buffer layer for organic and perovskite solar cells. Organic Electronics. 2016;**38**:150-157

[11] Bu IY-Y, Chen S. Improved crystalline silicon solar cells by light harvesting zinc oxide nanowire arrays. Optik: International Journal for Light and Electron Optics. 2016;**127**:10355-10359

[12] Mamat MH, Malek MF, Hafizah NN, Asiah MN, Suriani AB, Mohamed A, et al. Effect of oxygen flow rate on the ultraviolet sensing properties of zinc oxide nanocolumn arrays grown by radio frequency magnetron sputtering. Ceramics International. 2016;**42**:4107-4119

[13] Modaresinezhad E, Darbari S. Realization of a room-temperature/self-powered humidity sensor, based on ZnO nanosheets. Sensors and Actuators B: Chemical. 2016;**237**:358-366

[14] Ye Z, Wang T, Wu S, Ji X, Zhang Q. Na-doped ZnO nanorods fabricated by chemical vapor deposition and their optoelectrical properties. Journal of Alloys and Compounds. 2017;**690**:189-194

[15] Wang B, Duan Y, Zhang J. A controllable interface performance through varying ZnO nanowires dimensions on the carbon fibers. Applied Surface Science. 2016;**389**:96-102

[16] Zhou S-L, Zhang S, Liu F, Liu J-J, Xue J-J, Yang D-J, et al. ZnO nanoflowers photocatalysis of norfloxacin: Effect of triangular silver nanoplates and water matrix on degradation rates. Journal of Photochemistry and Photobiology A: Chemistry. 2016;**328**:97-104

[17] Schöttle C, Feldmann C. ZnO hollow nanospheres via Laux-like oxidation of Zn0 nanoparticles. Solid State Sciences. 2016;**55**:125-129

[18] Lin CJ, Liao S-J, Kao L-C, Liou SYH. Photoelectrocatalytic activity of a hydrothermally grown branched Zno nanorod-array electrode for paracetamol degradation. Journal of Hazardous Materials. 2015;**291**:9-17

[19] Dou Y, Wu F, Mao C, Fang L, Guo S, Zhou M. Enhanced photovoltaic performance of ZnO nanorod-based dye-sensitized solar cells by using Ga doped ZnO seed layer. Journal of Alloys and Compounds. 2015;**633**:408-414

[20] Ibrahim AA, Umar A, Kumar R, Kim SH, Bumajdad A, Baskoutas S. Sm_2O_3-doped ZnO beech fern hierarchical structures for nitroaniline chemical sensor. Ceramics International. 2016;**42**:16505-16511

[21] Bhatia S, Verma N, Bedi RK. Optical application of Er-doped ZnO nanoparticles for photodegradation of direct red-31 dye. Optical Materials. 2016;**62**:392-398

[22] Xu K, Liu C, Chen R, Fang X, Wu X, Liu J. Structural and room temperature ferromagnetic properties of Ni doped ZnO nanoparticles via low-temperature hydrothermal method. Physica B: Condensed Matter. 2016;**502**:155-159

[23] Manivannan A, Dutta P, Glaspell G, Seehra MS. Nature of magnetism in Co and Mn substituted ZnO prepared by sol-gel technique. British Journal of Applied Physics. 2006;**99**:08M110 (3 pages)

[24] Moditswe C, Muiva CM, Juma A. Highly conductive and transparent Ga-doped ZnO thin films deposited by chemical spray pyrolysis. Optik: International Journal for Light and Electron Optics. 2016;**127**:8317-8325

[25] Ismail AS, Mamat MH, Sin NDMd, Malek MF, Zoolfakar AS, Suriani AB, et al. "Fabri cation of hierarchical Sn-doped ZnO nanorod arrays through sonicated sol-gel immersion for room temperature, resistive-type humidity sensor applications. Ceramics International. 2016;**42**:9785-9795

[26] Dung ND, Son CT, Loc PV, Cuong NH, Kien PT, Huy PT, et al. Magnetic properties of sol-gel synthesized C-doped ZnO nanoparticles. Journal of Alloys and Compounds. 2016;**668**:87-90

[27] Ravichandran K, Dineshbabu N, Arun T, Manivasaham A, Sindhuja E. Synergistic effects of Mo and F doping on the quality factor of ZnO thin films prepared by a fully automated home-made nebulizer spray technique. Applied Surface Science. 2017;**392**:624-633

[28] Salah N, Hameed A, Aslam M, Abdel-wahab MS, Babkair SS, Bahabri FS. Flow controlled fabrication of N doped ZnO thin films and estimation of their performance for sunlight photocatalytic decontamination of water. Chemical Engineering Journal. 2016;**291**:115-127

[29] Meshram SP, Adhyapak PV, Amalnerkar DP, Mulla IS. Cu doped ZnO microballs as effective sunlight driven photocatalyst. Ceramics International. 2016;**42**:7482-7489

[30] Kim S-K, Gopi CVVM, Srinivasa Rao S, Punnoose D, Kim H-J. Highly efficient yttrium-doped ZnO nanorods for quantum dot-sensitized solar cells. Applied Surface Science. 2016;**365**:136-142

[31] Kim DH, Park J-H, Lee TI, Myoung J-M. Superhydrophobic Al-doped ZnO nanorods-based electrically conductive and self-cleanable antireflecting window layer for thin film solar cell. Solar Energy Materials and Solar Cells. 2016;**150**:65-70

[32] Meshki M, Behpour M, Masoum S. Application of Fe doped ZnO nanorods-based modified sensor for determination of sulfamethoxazole and sulfamethizole using chemometric methods in voltammetric studies. Journal of Electroanalytical Chemistry. 2015;**740**:1-7

[33] Anbia M, Fard SEM. Humidity sensing properties of Ce-doped nanoporous ZnO thin film prepared by sol-gel method. Journal of Rare Earths. 2012;**30**:38-42

[34] Peng X, Chu J, Yang B, Feng PX. Mn-doped zinc oxide nanopowders for humidity sensors. Sensors and Actuators B: Chemical. 2012;**174**:258-262

[35] Lee S-W, Choi BI, Kim JC, Woo S-B, Kim Y-G, Kwon S, et al. Sorption/desorption hysteresis of thin-film humidity sensors based on graphene oxide and its derivative. Sensors and Actuators B: Chemical. 2016;**237**:575-580

[36] Yao Y, Xue Y. Impedance analysis of quartz crystal microbalance humidity sensors based on nanodiamond/graphene oxide nanocomposite film. Sensors and Actuators B: Chemical. 2015;**211**:52-58

[37] Churenkov AV. Resonant micromechanical fiber optic sensor of relative humidity. Measurement. 2014;**55**:33-38

[38] Burgmair M, Zimmer M, Eisele I. Humidity and temperature compensation in work function gas sensor FETs. Sensors and Actuators B: Chemical. 2003;**93**:271-275

[39] Yao Y, Chen X, Li X, Chen X, Li N. Investigation of the stability of QCM humidity sensor using graphene oxide as sensing films. Sensors and Actuators B: Chemical. 2014;**191**:779-783

[40] Tang Y, Li Z, Ma J, Wang L, Yang J, Du B, et al. Highly sensitive surface acoustic wave (SAW) humidity sensors based on sol-gel SiO_2 films: Investigations on the sensing property and mechanism. Sensors and Actuators B: Chemical. 2015;**215**:283-291

[41] Ruiz V, Fernández I, Carrasco P, Cabañero G, Grande HJ, Herrán J. Graphene quantum dots as a novel sensing material for low-cost resistive and fast-response humidity sensors. Sensors and Actuators B: Chemical. 2015;**218**:73-77

[42] Hendi AA, Alorainy RH, Yakuphanoglu F. Humidity sensing characteristics of Sn doped Zinc oxide based quartz crystal microbalance sensors. Journal of Sol-Gel Science and Technology. 2014;**72**:559-564

[43] Zhao T, Fu Y, Zhao Y, Xing L, Xue X. Ga-doped ZnO nanowire nanogenerator as self-powered/active humidity sensor with high sensitivity and fast response. Journal of Alloys and Compounds. 2015;**648**:571-576

[44] Hsu N-F, Chang M, Hsu K-T. Rapid synthesis of ZnO dandelion-like nanostructures and their applications in humidity sensing and photocatalysis. Materials Science in Semiconductor Processing. 2014;**21**:200-205

[45] Hou J-L, Wu C-H, Hsueh T-J. Self-biased ZnO nanowire humidity sensor vertically integrated on triple junction solar cell. Sensors and Actuators B: Chemical. 2014;**197**:137-141

[46] Hong H-S, Chung G-S. Controllable growth of oriented ZnO nanorods using Ga-doped seed layers and surface acoustic wave humidity sensor. Sensors and Actuators B: Chemical. 2014;**195**:446-451

[47] Li Q, Liu X, Gu M, Huang S, Zhang J, Liu B, et al. Enhanced X-ray excited luminescence of Ga- and In-doped ZnO nanorods by hydrogen annealing. Materials Research Bulletin. 2017;**86**:173-177

[48] Son NT, Noh J-S, Park S. Role of ZnO thin film in the vertically aligned growth of ZnO nanorods by chemical bath deposition. Applied Surface Science. 2016;**379**:440-445

[49] Mamat MH, Sahdan MZ, Khusaimi Z, Ahmed AZ, Abdullah S, Rusop M. Influence of doping concentrations on the aluminum doped zinc oxide thin films properties for ultra-violet photoconductive sensor applications. Optical Materials. 2010;**32**:696-699

[50] Yue S, Wang L, Zhang D, Yang S, Guo X, Lu Y, et al. Facile synthesis of one-dimensional Fe-doped ZnO nanostructures from a single-source inorganic precursor. Materials Letters. 2014;**135**:107-109

[51] Malek MF, Mamat MH, Khusaimi Z, Sahdan MZ, Musa MZ, Zainun AR, et al. Sonicated sol–gel preparation of nanoparticulate ZnO thin films with various deposition speeds: The highly preferred c-axis (0 0 2) orientation enhances the final properties. Journal of Alloys and Compounds. 2014;**582**:12-21

[52] Malek MF, Mamat MH, Musa MZ, Soga T, Rahman SA, Alrokayan SAH, et al. Metamor phosis of strain/stress on optical band gap energy of ZAO thin films via manipulation of thermal annealing process. Journal of Luminescence. 2015;**160**:165-175

[53] Pascariu P, Airinei A, Olaru N, Petrila I, Nica V, Sacarescu L, et al. Microstructure, electrical and humidity sensor properties of electrospun NiO–SnO_2 nanofibers. Sensors and Actuators B: Chemical. 2016;**222:**1024-1031

[54] Khuili M, Fazouan N, El Makarim HA, El Halani G, Atmani EH. Comparative first principles study of ZnO doped with group III elements. Journal of Alloys and Compounds. 2016;**688**(Part B): 368-375

[55] Bai S, Guo T, Zhao Y, Sun J, Li D, Chen A, et al. Sensing performance and mechanism of Fe-doped ZnO microflowers. Sensors and Actuators B: Chemical. 2014;**195**:657-666

[56] Gao F, Liu XY, Zheng LY, Li MX, Bai YM, Xie J. Microstructure and optical properties of Fe-doped ZnO thin films prepared by DC magnetron sputtering. Journal of Crystal Growth. 2013;**371**:126-129

[57] Gürbüz O, Kurt İ, Çalışkan S, Güner S, Influence of Al concentration and annealing temperature on structural, optical, and electrical properties of Al co-doped ZnO thin films. Applied Surface Science. 2015;**349**:549-560

[58] Musat V, Teixeira B, Fortunato E, Monteiro RCC, Vilarinho P. Al-doped ZnO thin films by sol–gel method. Surface and Coatings Technology. 2004;**180-181:**659-662

[59] Wang W, Li Z, Liu L, Zhang H, Zheng W, Wang Y, et al. Humidity sensor based on LiCl-doped ZnO electrospun nanofibers. Sensors and Actuators B: Chemical. 2009;**141**:404-409

[60] Md Sin ND, Mamat MH, Malek MF, Rusop M. Fabrication of nanocubic ZnO/ SnO_2 film-based humidity sensor with high sensitivity by ultrasonic-assisted solution growth method at different Zn:Sn precursor ratios. Applied Nanoscience. 2014;**4**:829-838

[61] Yang CC, Shen JY. Well-defined sensing property of ZnO:Al relative humidity sensor with selected buffer layer. Vacuum. 2015;**118**:118-124

[62] Zhu D, Hu T, Zhao Y, Zang W, Xing L, Xue X. High-performance self-powered/active humidity sensing of Fe-doped ZnO nanoarray nanogenerator. Sensors and Actuators B: Chemical. 2015;**213**:382-389

Chapter 10

Performance Evaluation of Nanostructured Solar Cells

Abouelmaaty M. Aly

Additional information is available at the end of the chapter

http://dx.doi.org/10.5772/67405

Abstract

Nanotechnology is making great contributions in various fields including harvesting solar energy through solar cells since nanostructured solar cells can provide high performance with lower fabrication costs. The transition from fossil fuel energy to renewable sustainable energy represents a major technological challenge for the world. The solar cells industry has grown rapidly in recent years due to strong interest in renewable energy in order to handle the problem of global climate change that is now believed to occur due to the use of fossil fuels. Cost is an important factor in the eventual success of any solar technology since inexpensive solar cells are needed to provide electricity especially for rural areas and for underdeveloped countries. Therefore, new developments in nanotechnology may open the door to the production of cheaper and more efficient solar cells by reducing the manufacturing costs of solar cells. This chapter covers a review of the progress that has been made to date to enhance efficiencies of various nanostructures used in solar cells including utilizations of all the wavelengths present in of the solar spectrum.

Keywords: nanostructured solar cells, power efficiency, quantum wells, quantum wires, quantum dots, intermediate bands, solar concentrations

1. Introduction

The worldwide demand for energy has an environmental impact on the global climate since most of the energy in recent decades has come from the combustion of fossil fuels. Many technologies are being considered to supplement this energy such as renewable energy sources (RESs). RESs are the key to long-term solution of industrialized economies from complete dependence on fossil fuels [1]. Solar energy conversion is the most attractive source of RESs because solar energy is abundant and it is free. In the past 50 years, silicon (Si) solar cells (SCs) have been developed with demonstrated efficiency of nearly 25%, which is very close to the theoretical limit for a single junction under one sun concentration of ~31% [2]. Unfortunately, this type of bulk Si SCs is too expensive for mass production. For the solar energy conversion, Si is not the ideal material because the indirect bandgap of Si has a significant negative impact on the optical absorption [1]. As a result, the 90% of sunlight is absorbed when the energy of photons is greater than the bandgap of Si and the thickness of Si is around 125 μm, whereas for a direct bandgap material such as gallium arsenide (GaAs), the required thickness is around 0.9 μm [3]. In addition, the incoming photons on the bulk Si SCs should have the same energy of the bandgap of Si to obtain a free electron. The photon will have dissipated inside the bulk Si, if it has energy less than bandgap of Si. If it has more energy than bandgap of Si, the extra energy will be lost as heat. These factors cause the loss of around 70% of the radiation energy incident on the cell [4]. This trade-off between the two factors is evident from the current density-voltage (*J*- *V*) curves shown in **Figure 1(a)** for some elemental semiconductors, such as germanium (Ge) or Si, and binary compounds, such as GaAs, aluminium antimonide (AlSb) and gallium phosphide (GaP). The main problem of the bulk SCs is evident in **Figure 1(a)** in that when the output current increases the output voltage decreases and vice versa based on bandgap energies of these materials. Therefore, the optimum bandgap energy of SC matches the maximum efficiency can be calculated as shown in **Figure 1(b)**. This figure shows the maximum efficiency versus bandgap energy of bulk SCs under Blackbody spectrum and more realistic AM0 and AM1.5 solar spectra at one sun and maximum concentrations.

Nanotechnology is an important vehicle to reduce the SC cost and to improve its performance by utilizing nanostructured materials in SCs [6]. Nanostructured materials are generally defined as those materials whose structural elements (crystallites or molecules) have dimensions in the 1 –100 nm range. Attention of both academic and industrial researchers for these materials over the past decade arises from the remarkable differences in fundamental electrical, optical and magnetic properties that occur with change in size while controlling the construction of the materials at the atomic level. The motivation for using nanostructured materials emerges from their specific physical and chemical properties. Enhancing the regular crystalline structure using nanocrystalline materials in the form of thin films or multi-layers SCs can increase the absorbance of all incident solar spectra [7, 8]. In nanoparticle-based SCs, the particles should be sufficiently close to one another to transfer the charges directly. Recently, significant progress has

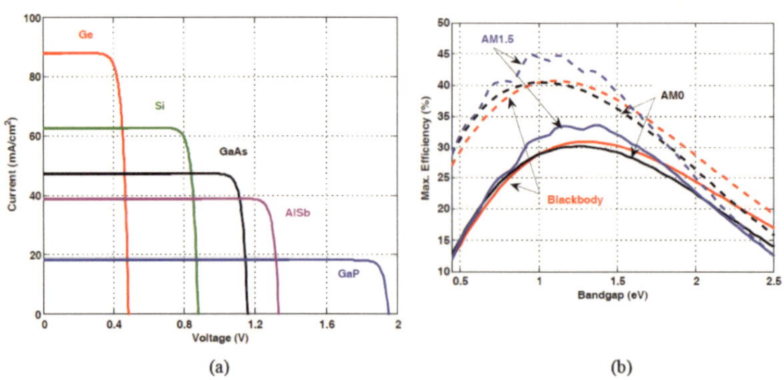

Figure 1. (a) *J-V* curves of some semiconductors for bulk SCs and (b) maximum efficiencies versus bandgap energies of bulk SCs with different spectra at one sun (sold lines) and maximum concentration (dashed lines) [5].

been made in improving the overall efficiencies of SC structures, including the incorporation of potential well and nanocrystalline materials. The present investigation deals with some potential applications of nanostructured materials in solar energy conversion and to give an overview on current research topics in this field. This chapter includes the following sections: preparation of SCs in Section 2; fundamental of nanostructure confinement in Section 3; nanostructures for SC applications in Section 4 and conclusions in Section 5.

2. Preparation of solar cells

Over the past 20 years, electrical power generation output of SCs has grown up to 5×10^9 W (5 GW). This is still small in comparison to the world total electric generation capacity of 4×10^{12} W (4 TW) although it represents a large step forward in this promising renewable energy technology [9]. If the present rate of growth continues, solar energy could become the dominating power generation method by the end of this century. The first generation of SCs comprises technologies where the photon absorber is a thin *p*-doped single crystal Si wafer. Employing Si in the form of monocrystalline or polycrystalline material for this purpose is understandable in view of its abundance, stability, non-toxicity and decades of industry experience. However, single-crystal Si wafers are very expensive to produce due to the demands of high purity and high accuracy of sawing a single wafer from Si. The proposed practical bound on cell efficiency under one sun has been estimated to be about 25%. The second generation of SCs aims at reducing the costs of producing thin-film SCs by growing thin layers of Si and other semiconductors on glass substrates. This generation of SCs uses materials, such as cadmium telluride (CdTe), copper-indium-gallium selenide (CIGS), copper-indium sulphide (CIS) and amorphous Si (a-Si). These materials are much cheaper than using a single-crystal Si but have the downside of leading to less efficient SCs with efficiency around 16% due to structural defects [10]. The third generation of SCs is made from variety of new materials besides Si to make solar energy more efficient over a wider band of solar energy including IR band. There are several technologies in this generation based on nanotechnology such as hot carrier, tandem or multi-junction, and intermediate-band SCs. The best overall efficiency for this generation has reached to 44.7% at one sun. Research into the fourth generation of SCs has recently begun, and they are made from hybrid organic materials, which are low cost and nanostructure inorganic materials with stable lifetime. This technology could significantly improve the harvesting and conversion of solar energy [11]. This generation of materials could improve the efficiency while maintaining their low cost when compared to third-generation materials.

3. Fundamentals of nanostructure confinement

This section gives an overview of the optical properties of quantum-confined semiconductor structures. The potential barriers in these artificial structures can confine the motion of electrons and holes (charge carriers, CCs) in one or more directions [12]. The optical properties of bulk solids do not usually depend on their size. For example, ruby crystals have the same red colour regardless of how big they are [13]. This statement is only true as long as the dimensions of the crystal are large. For very small crystals, the optical properties are dependent on the size such as semiconductor-doped glasses. These contain very small semiconductor micro-crystals within a colourless glass, and the colour of the filter can be altered just by changing the size of the crystals. The amount of dependence of optical properties in very small crystals is the result of the quantum confinement effect. The bulk crystal will not exhibit any quantum size effects. To observe quantum size effects, the layers should be thin. The general scheme for classifying quantum-confined structures is given in **Table 1**. The artificial structures are classified as to whether the CCs are free or confined in one, two or three dimensions. For bulk semiconductors, the CCs are free (not confined) to move within their respective bands in all the three directions. For quantum-confined semiconductors, the structures are respectively called quantum wells (QWs) for confinement in one-dimensional (1D), quantum wires (QRs) for confinement in two-dimensional (2D) and quantum dots (QDs) for confinement in three dimensions (3D). The nanocrystal dimensions required to observe quantum-confinement effect, have to be produced by the advanced techniques such as advanced epitaxial crystal growth (QW structures), epitaxial growth on patterned substrates (QR structures) and spontaneous growth technique (QDs structures) [12].

In a nanocrystalline particle with dimensions L_x, L_y and L_z, the CCs are localized in the region with the minimum potential energy in the nanoparticle, and its energy spectrum is no longer formed from allowed and forbidden bands as in crystals, but from discrete levels as shown in **Figure 2(a)** and given by [14]:

$$E(k_x,\ k_y, k_z) = E_c + \frac{\hbar^2}{2m}\left(\frac{p\pi}{L_x}\right)^2 + \frac{\hbar^2}{2m}\left(\frac{q\pi}{L_y}\right)^2 + \frac{\hbar^2}{2m}\left(\frac{r\pi}{L_z}\right)^2 = E_{pqr} \tag{1}$$

Here ħ is the Plank's constant, m is the effective mass, E is the total energy of CCs, k is the wave-vector, E_c is the edge energy of CCs in the conduction band (CB) or valence band (VB), and p, q and r are integer numbers. These discrete levels result from the Schrödinger Equation satisfied by the CCs in the materials [15]. A discrete energy spectrum is similar to that in atoms or molecules. A nanoparticle is called QD if the discrete energy levels can be observed, that is, if the difference between adjacent discrete energy levels is higher than the thermal vibration energy, $k_B T$ as T is the temperature and k_B is the Boltzmann constant. This implies that $\hbar^2\pi^2/2mL_x^2$, $\hbar^2\pi^2/2mL_y^2$ and $\hbar^2\pi^2/2mL_z^2$ are higher than $k_B T$ or $L_x, L_y, L_z < \sqrt{\hbar^2\pi^2/2mk_B T}$.

In a QD, the CCs are not free to move because it is localized in the dot. There are also structures in which CCs movement is allowed along a single direction, that is, QRs, or in one plane, that is, QWs. Unlike these structures, in a bulk crystalline material the CCs can freely move along all three spatial directions and the energy dependence on the k and it is given by

$$E(k) = E_c + \hbar^2 k^2/2m = E_c + \hbar^2(k_x^2 + k_y^2 + k_z^2)/2m \tag{2}$$

Figure 2(b) shows a QR with L_y and L_z dimensions along the confinement directions. The CCs can freely move along only the x-direction and the dispersion relation is expressed as

$$E(k_x,\ k_y, k_z) = E_c + \frac{\hbar^2}{2m}\left(\frac{q\pi}{L_y}\right)^2 + \frac{\hbar^2}{2m}\left(\frac{r\pi}{L_z}\right)^2 + \frac{\hbar^2 k_x^2}{2m} = E_{s,qr} + \frac{\hbar^2 k_x^2}{2m} \tag{3}$$

http://dx.doi.org/10.5772/67405

Structure	Quantum confinement	Dimensionality
Bulk	None	3
QW	1D	2
QR	2D	1
QD	3D	0

Table 1. Classification of quantum-confined structures.

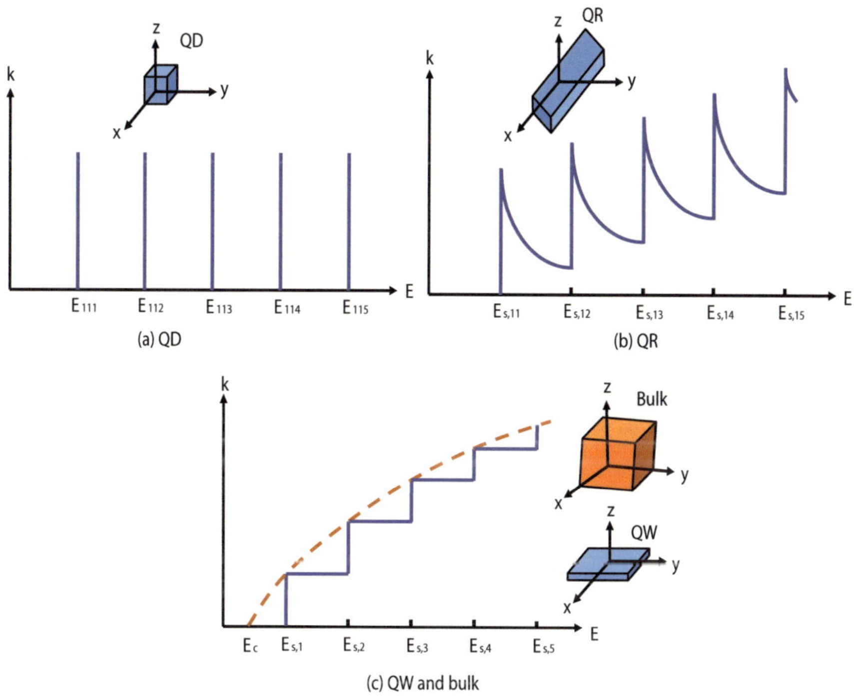

Figure 2. *E-k* relations for QD, QR, and QW and bulk crystalline lattices. (a) QD,(b) QR and (c) QW and bulk.

As in a QD, the difference between adjacent discrete levels must be higher than $k_B T$,that is, $L_y,\ L_z < \sqrt{\hbar^2 \pi^2/2m\,k_B T}$.

Analogously, in a QW as shown in **Figure 2(c)**, the CCs are free to move along the x and y directions and the dispersion relation is

$$E\left(k_x,\ k_y, k_z\right) = E_c + \frac{\hbar^2}{2m}\left(\frac{r\pi}{L_z}\right)^2 + \frac{\hbar^2(k_x^2 + k_y^2)}{2m} = E_{s,r} + \frac{\hbar^2(k_x^2 + k_y^2)}{2m} \tag{4}$$

These discrete energy levels can be observed if the width of the QW satisfies the relation $L_z < \sqrt{\hbar^2 \pi^2/2m\,k_B T}$. The dashed line in **Figure 2(c)** represents the density of k versus E in a bulk crystalline lattice as the CCs freely move in three dimensions and is given by the equation $E = \hbar^2 k^2/2m$.

In QWs, QRs and QDs, the allowed region in which the CCs move is called *potential well*, and the region where this motion is forbidden is referred to as *potential barrier*. The excitation of an electron from the VB to CB as a result of photon absorption is only possible if the photon energies (E) are higher than the bandgap energy (E_g) of the material that contains the QW (for example); this minimum photon energy also depends on the width of the well and the height of the potential barrier as shown in **Figure 3(a)**.

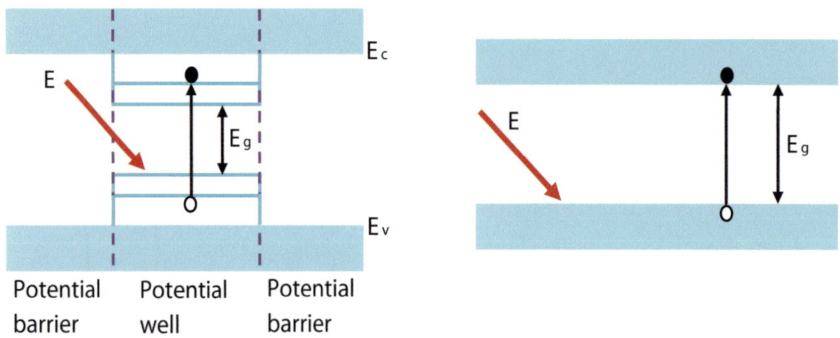

Figure 3. Construction of energy band diagram for two semiconductor material.

More precisely, the electron transition from the VB with upper edge (E_v) to the CB with bottom edge (E_c) occurs between the corresponding discrete levels in the *potential well*. In bulk crystals, the similar electron excitation occurs if the incident photon energy E is equal to E_g, as shown in **Figure 3(b)**. The absorption spectrum depends on the nanoparticle diameter due to the above relations between the CC energy and the dimensions of the *potential well*. When several QWs, QRs or QDs are separated by small distances, the CCs can jump from one structure to another if their energy (thermal) is high enough to overcome the potential barrier.

4. Nanostructures for solar cell applications

With improved possibilities for the control of material texture on the nm scale, nanostructured SCs have received increased scientific attention in recent years. Nanostructures can allow efficient SCs to be made from cheaper materials, such as Si and titanium dioxide (TiO_2) [16]. Although there will be cost barriers involved in developing mass production techniques for nanostructured SCs, the use of cheaper raw materials will allow a cost reduction of commercial SCs. Nanomaterials and nanostructures hold promising possibilities to enhance the performance of SCs by improving both optical absorption and photo-carrier collection. Meanwhile, the new materials and structures can be fabricated in a low-cost fashion, enabling cost-effective production of SCs. As the performance of SCs largely depends on the above two factors, they have to be optimized for SCs with suitable energy conversion efficiency. Nevertheless, the requirements in optimizing these factors can be in conflict. For example, in a planar-structured SC thicker materials are needed in order to achieve sufficient optical absorption; however, it will reduce carrier collection probability due to the increased minority carrier diffusion path length and vice versa. In fact, recent studies have shown that nanostructures not only improve optical absorption by utilizing the light trapping effect but also facilitate the photocarrier collection through perpendicular directions of light propagation and carrier collection. The nanostructures discussed in the following sections are classified into four types: nanocomposites, QWs, QRs and QDs.

4.1. Nanocomposites materials for solar cells

Nanocomposites and nanostructured materials are now being investigated for their potential applications in SCs. In nanoparticles of diameter, *d*, the number of atoms residing on their surfaces varies as 1/*d* and hence they can become significant fractions of the atoms present in the core of the few nm size nanoparticles. In such a case, the surface interactions control the behaviour of nanoparticles. Therefore, these small particles often have different characteristics and properties than larger pieces of the same material. Nanostructured layers in thin film SCs offer a number of important advantages [17]: (i) due to multiple reflections, the effective optical path for absorption is much larger than the actual film thickness and (ii) light-generated CCs need to travel over a much shorter path and thus recombination losses are greatly reduced. As a result, the absorber layer thickness in nanostructured SCs can be as thin as 150 nm instead of several micrometres in the conventional thin film SCs and thus reduced installation costs achieved; (iii) the energy bandgap of various layers can be designed to the desired value by varying the size of nanoparticles and (iv) reduced manufacturing costs as a result of using a low-temperature process instead of the high-temperature vacuum deposition process for conventional SCs. Thin films of polycrystalline CdTe, cadmium selenide (CdSe) and cadmium sulphide (CdS) have been reported as the most promising photovoltaic materials for thin film SCs [16]. Other types of nanostructured materials used for SC applications are the hot carrier SCs (HCSCs) and the dye-sensitized SCs (DSSCs).

In a HCSC, the rate of photo-excited carrier is slow enough to allow time for the carriers to be collected and thus allowing higher voltages to be achieved from the cell [18]. The bulk cell is designed to collect the CCs before the hole and the electron recombined while the hot carrier cell catches CCs before the carrier cooling stage. For the HCSC to be effective, CCs should be collected from the absorber over a very small energy range. Hence, special contacts are used to prevent the contacts from cooling the carriers. The limiting efficiency of this approach can reach 86.8%, which is same as an infinite tandem cell stack. However, in order to achieve this limiting efficiency, carrier cooling rates should be reduced or radiative recombination rates should be accelerated [19].

Developing a hot carrier absorber material, which exhibits sufficiently slow carrier cooling to maintain a hot carrier population under realistic levels of sun concentration, is a key challenge. A candidate for the absorber material is a QD super-lattice [20]. InGaAs/gallium arsenide phosphide (GaAsP) is proposed as a suitable absorber material and the GaAs surface buffer layer was reduced in thickness to maximize photon absorption in the well region. An enhanced hot carrier effect was observed in the optimized structures. The HCSC with indium nitride (InN) absorber layer gives a highest efficiency of 52% as shown in **Figure 4** [21]. The efficiency of the HCSC, with

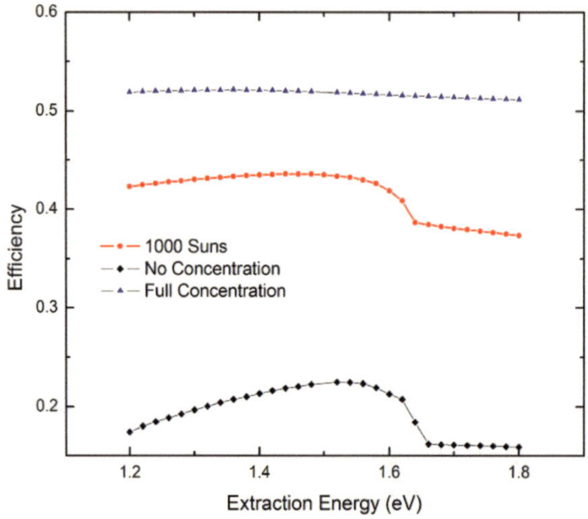

Figure 4. HCSC efficiency versus carrier extraction energy with absorber layer (InN) thickness = 50 nm at different sun concentrations [21].

gallium antimonide (GaSb)-based heterostructures as absorber candidates, is improved significantly compared to a fully thermalized single *p-n* junction of similar bandgap energy.

A DSSC is a type of photoelectrochemical SC [22]. In a DSSC, dye molecules are used to sensitize wide-bandgap energy semiconductors (~3.2 eV), such as TiO_2 and ZnO, which assist in separating electrons from photo-excited dye molecules. These materials are very inexpensive compared to expensive Si or III-V group semiconductors. As a result, much cheaper solar energy at \$1 or less per peak Watt (\$1/pW) can be achieved [23]. The DSSC involves a set of different layers of components stacked in serial. Incoming photons excite electrons in the dye which are efficiently captured by the TiO_2 and transported to the anode [24]. On the other side, an electrolyte (I^-/I_3^-) replenishes the electrons in the dye (i.e. $3I^- = I_3^- + 2e^-$), and thus completes the flow of current through the cathode. In order to maximize the contact area between the wide-bandgap energy semiconductors and the dye, it is advantageous to have the wide-bandgap energy semiconductors in the form of nanoparticles or nanotubes. In combination with new absorbing molecules, such structures have helped to improve the performance of DSSCs.

Figure 5 uses carbon nanotube (CNT) fibres, which have been used as a conductive material to support the dye-impregnated TiO_2 nanoparticles. Since TiO_2 nanoparticles present in CNTs are capable of injecting electrons from their excited state and therefore separation and flow of the charges are improved. CNTs are also popular materials for DSSC counter electrode (cathode) fabrication. In addition to enhance conversion efficiency, the cells with CNT counter electrode are expected to provide several advantages including nanoscale conducting channels, lightweight and low cost, as well as improved mechanical properties and thermal stability [25]. Since the invention of the DSSCs in 1990, efficiency has improved from ~2% to ~12%. It has a major cost advantage over bulk Si SCs. Some of the main drawbacks of DSSCs are the use of liquid electrolyte and dyes, long-term lifetime and degradation [26]. Most DSSCs use the organic dyes, which have large optical absorption in the visible range

http://dx.doi.org/10.5772/67405

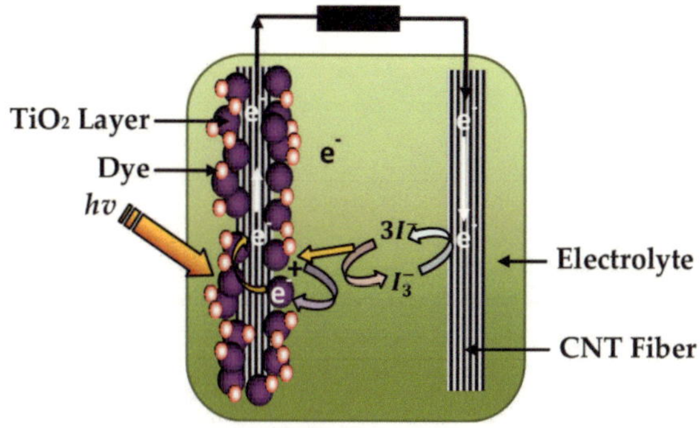

Figure 5. Wire-shaped DSSC made from two CNT fibres.

of 300–700 nm but weak absorption in the IR spectrum. Searching for new materials with absorption spectra extending into the IR range is decisive for enhancing the performance of SCs. A promising alternative for broadband solar absorbers will be inorganic semiconductor sensitizers. In 2013, the performance of semiconductor-sensitized SCs (SSSCs) was improved and the efficiency has increased from 12% to 15% [23]. Inorganic semiconductor sensitizers have several advantages over organic dyes such as tunable absorption bands due to the quantum-size effect and multi-electron-hole pair generation by a single incident photon.

4.2. Quantum well solar cells

Since several decades, QWs have been studied for the application in electronics and photonics. They confine the CCs in 1D and create a sheet of CCs with well-defined energy levels and high mobility due to adjustment in the band structure [1]. Recently, concerted efforts for the manufacture of SCs by incorporating multi-QWs (MQWs) with lower bandgap energy in the active region of the device were carried out. In the beginning, MQWs depended on the III-V semiconductor materials, mainly GaAs and related alloys such as AlGaAs and InGaAs [27]. The main expected benefit of such SCs is high short circuit current due to the enhanced absorption. **Figure 6** shows a structure of GaAs *p-i-n* SC with a single InGaAs QW in intrinsic region [28]. In this work, a short circuit current density is an enhancement, as the absorption in QWs is very high due to the carrier density obtained by quantum confinement in the plane of the well. On the other hand, there is often a corresponding reduction in the open circuit voltage due to the inclusion of lower bandgap energy material which could be overcompensated through the increase in short circuit current from the QWs. GaAs SCs currently hold the world efficiency record for single junction SCs. MQW solar cells (MQWSCs) can achieve optimal bandgap energies for the highest single-junction efficiencies due to the tunability of the QW width and composition. However, the increase in the number of QWs causes mismatch in the lattice and, therefore, disorder occurs in the open circuit voltage [29]. In reference [28], strain-balanced GaAsP/InGaAs MQW in i-region has been studied. The GaAsP/InGaAs MQW strain-balanced SC (SB-QWSC) has shown an extraordinary performance for the MQW cell design, achieving high efficiency. The dependence of conversion efficiency on indium (In) and phosphide (P) compositions is examined in **Figure 7** for 20 layers of QWs.

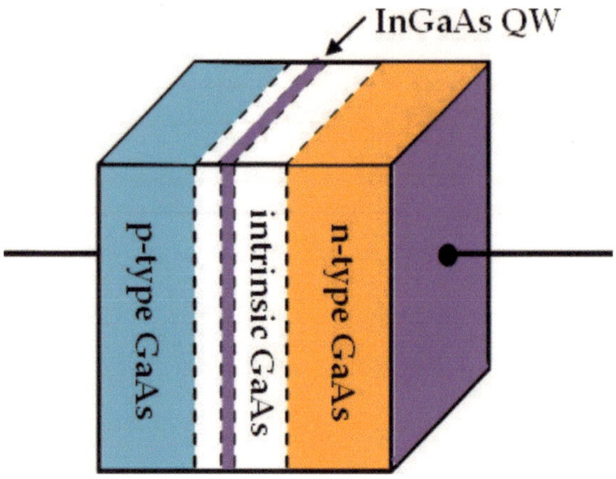

Figure 6. GaAs *p-i-n* SC with a single InGaAs QW in *i*-region.

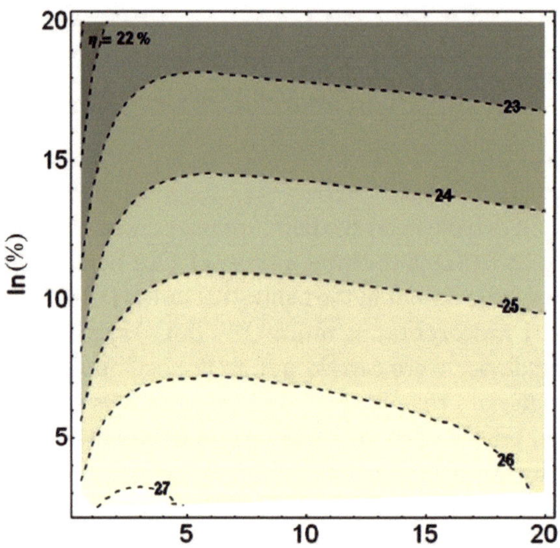

Figure 7. Contour plot for conversion efficiency (η) versus In and P compositions for 20 layers of QWs and each with thickness of 15 nm [28].

Each well has thickness of 15 nm. In and P compositions are varied in the range from 0.5% to 20% during the development process of the contour plot. It is evident that the highest efficiency equals 27%. This occurs at a composition of not greater than 3% and 5%, respectively, for In and P.

4.3. Quantum wires solar cells

A number of researchers proposed to use QRs in all inorganic structures to produce SCs with both improved efficiency as well as greater stability compared to organic SCs [30, 31]. These structures include vertical arrays of radial or axial *p-n* junction QRs to confine the CCs in 2D and to enhance photon absorption, see **Figure 8**. The radial geometry offers a large area of the depletion region along the QRs and a small QRs volume combined with short carrier diffusion lengths before collection at the contacts as a usable current. The axial geometry offers greater flexibility with respect to materials combinations. Challenges for both architectures are overcoming the barrier of single to tandem junction performance and limitations of surface recombination which leads to loss of CCs.

The potential for improved performance and cost reductions of QR arrays over their bulk SCs is mainly due to (i) increased absorption due to diffuse light scattering in QR arrays; (ii) short collection lengths of minority carriers that are radially separated and (iii) flexibility of cell integration on a variety of low-cost carrier substrates [30]. Recently, the nanostructure of Si QRs has attracted great attention because of its excellent anti-reflection and light trapping effect [32]. This structure is a candidate to lower both the required quality and quantity of Si material. Many researchers have investigated Si QRSCs to provide support for further improvement. However, the efficiency of Si QRSCs still falls behind that of the bulk crystalline Si SCs as a result of the limitation of the extremely high surface recombination in Si QRs. The repression of carrier recombination in Si QRs turns out to be the primary focus for the performance improvement of Si QRSCs. Surface passivations, such as thermal oxidation, carbon thin films and chlorine dielectric treatment, have been widely studied to improve the electrical characteristics of Si QRSCs [33]. These techniques can only work on the recombination at the Si QRs surface. However, a recent attempt has demonstrated that, in addition to the surface recombination, high Auger recombination near the surface plays a key role in the limitation of the photogenerated carrier collection and cell efficiency in Si QRSCs [32]. The Auger recombination comes from the high doping related to in-diffusion through the large surface area of the Si nanostructure. This near surface Auger recombination grows over the surface recombination, especially for the excessive doping condition in Si QRSCs.

In order to reduce recombination both at and near the surface in Si QRs while maintaining good light trapping, an efficient method of Si nitride (SiN_x) passivation is proposed [34]. SiN_x can provide not only effective surface passivation but also bulk passivation because of the hydrogen diffusion which effectively reduces defect state density and suppresses the Auger recombination at and near the surface [35]. Also, to obtain the best recombination repression both at and near the

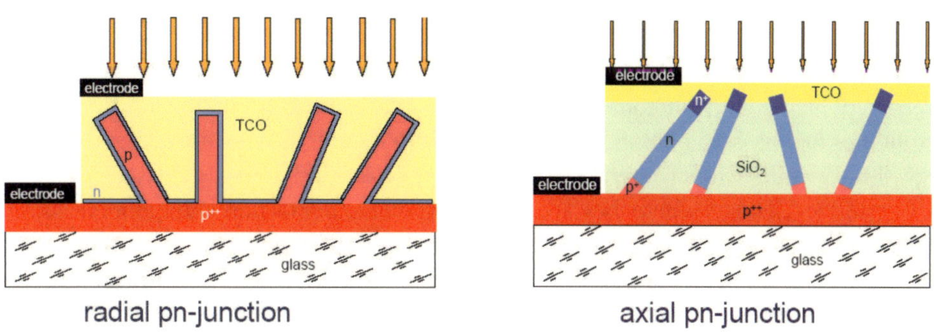

Figure 8. Two structures of QRSCs.

surface, another work proposed for surface passivation uses SiN_x combined with the Si dioxide (SiO_2). The SiO_2/SiN_x stack performs better due to the more homogeneous Si-SiO_2 self-oxidation interface as well as decreases surface state density [34]. The current-voltage characteristics of Si QRSC passivated with different materials that were measured under illumination conditions of AM1.5G are shown in **Figure 9**.

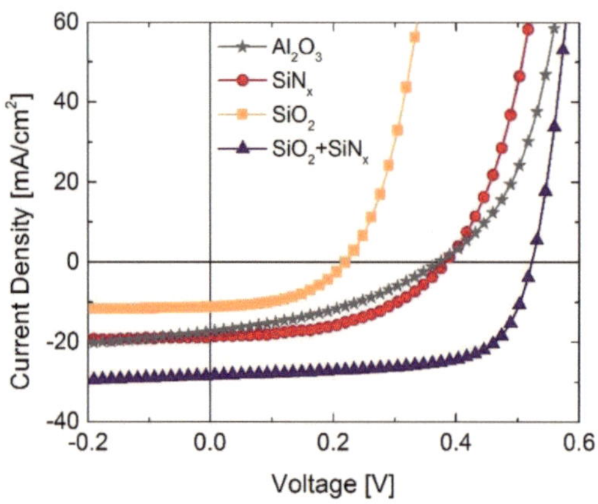

Figure 9. Current-voltage characteristics under AM1.5G illumination of the QR arrays covered with aluminium oxide (Al_2O_3), SiN_x, SiO_2 and a SiO_2/SiN_x stack [34].

Surface recombination on axial *p-n* junction QR arrays has a significant impact on both short circuit current density (J_{sc}) and open circuit voltage (V_{oc}). Through deep investigation of the effects of different passivations and Si QR lengths as well as effective control of carrier recombination, the efficiency has been recorded as 17.11% on large area (125 × 125 mm^2) Si QRSCs by conventional industrial manufacturing processes [34]. This idea opens a potential prospect for the practical fabrication of large size Si QRSCs with satisfactory conversion efficiency.

4.4. Quantum dot solar cells

QD structures have been implemented in various SC applications and it is possible to synthesize QDs in many compositions (semiconductors or metals) as well as coat them with dielectrics or additional semiconductors [1]. One of the early efforts in the utilization of QDs was the down conversion of high-energy photons [36]. The fundamental mechanism of down conversion is the absorption of high-energy photon with relaxation into intermediate states within the bandgap energy, thus emitting two lower energy photons. In a bulk SC, high-energy photons are excited well beyond the CB edge, where they are mostly lost due to interaction with phonons (thermalization). Down conversion aims to absorb these high-energy photons and shift them to lower energies that are matched to specific absorber material in the SC. Several schemes have been proposed to capture and convert the photons to lower energy that are better tuned to the bandgap energy of the semiconductor using nanostructures. One of them formed Si QDs in a dielectric layer deposited on top of a standard Si SC [37]. A small enhancement in quantum efficiency at short wavelengths was observed; however, the overall efficiency was not enhanced. The opposite of down conversion is the up conversion of photons of energy below the bandgap energy that are typically not absorbed by the bulk SC [38]. The typical mechanism of up conversion involves absorption of sub-bandgap energy light into an intermediate state, followed by further absorption of a second photon to the CB edge. Another novel band structure that can be obtained with QDs is the intermediate band (IB), which is indeed a form of up conversion. This is possible because the states of closely spaced QDs can overlap to form an effective band structure (min-bands) that when finely tuned yields an IB. In the mid-1990s, a new proposal called QD intermediate band SC (QDIBSC) appeared. In this type, the QDs are embedded in a *p-i-n* structure [39]. In addition, another proposal used multi-layered Si QD arrays to create an all Si-based multi-junction SC [40]. The following sub-sections include more details about multi-junctions and IB SCs based on QDs.

4.4.1. *Multi-junction solar cells*

Wavelengths of the light spectrum from UV to IR cannot be effectively captured by a semiconductor of particular E_g. If E_g is too large, most of the photons will not be absorbed; if it is too small, the photons will be absorbed, but much of their energy will be lost by thermalization. To overcome this phenomenon, various techniques have been suggested. One of these techniques is multi-junction SCs (MSCs). The concept of MSCs uses several *p-n* junction SCs with different bandgap energies in tandem which capture efficiently different parts of the light spectrum. The use of different materials in MSCs constitutes a technology challenge because it is very expensive such as GaInP/GaInAs/Ge SC which has achieved 41% efficiency [41]. It is good if one realizes different bandgap energies using only one material such as Si whose bandgap energy is varied using appropriate nanostructures. This idea is behind the all-Si multi-junction tandem SC using different size of QDs nanostructure. **Figure 10** shows this idea as the light comes to the top and the lowest wavelengths are first absorbed whereas the higher wavelengths go through and are absorbed in the middle and bottom cells. There are many ways of realizing such multi-layers. One of the best ways is to deposit alternate thin nanoscale layers of SiO_2 and a silicon-rich oxide (SiO_x, $x < 2$) [42]. During the annealing, the excess Si in the SiO_x layer precipitates to form Si nanocrystals separated by SiO_2 layers according to the reaction: $2\,SiO_x = (2-x)Si + SiO_2$, thus achieving the desired structure. The SiO_x layer thicknesses are controlled by varying the deposition time. It is well known that the ground state energies of CCs are raised by the quantum confinement, thus increasing the effective bandgap energy in a nanocrystal compared to the bulk. Further, the smaller the size of the nanocrystal, the larger the bandgap energy. **Figure 11(a)** and (b) shows the pictures of the multi-layer SiO_2/SiO_x structures, and the formation of Si nanocrystals after annealing at 900°C. Some bandgap energies of the nanocrystals were measured by using the photospectrometer. The results showed that the bandgap energy increases as the nanocrystal size decreases. When nanocrystal thickness changed from 2 to 10 nm, the bandgap energy changed from 2.5 to 1.45 eV, respectively [43]. This establishes that the multi-junction tandem SCs can be built with varying the bandgap energies by using only Si and its oxide.

4.4.2. *Intermediate band solar cells*

QDs super-lattice included in the active region of *p-i-n* single-junction SCs has been considered as one of the candidates to realize the QDIBSCs, as illustrated in **Figure 12**. The IBs allow for the absorption of low-energy photons that would otherwise be transmitted through the bulk

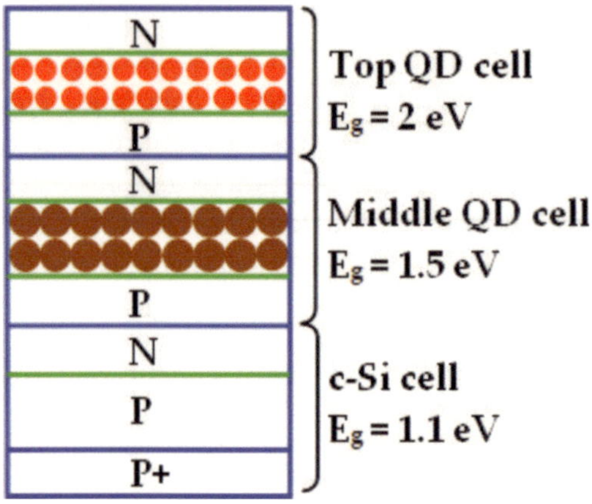

Figure 10. Three junctions "all-Si' tandem SC using two layers of Si QDs.

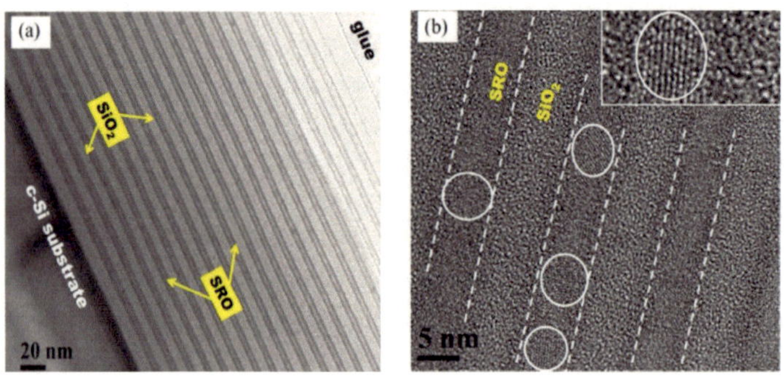

Figure 11. Pictures of 41-layers SiO_2/SiO_x structure before (a) and after (b) annealing [43].

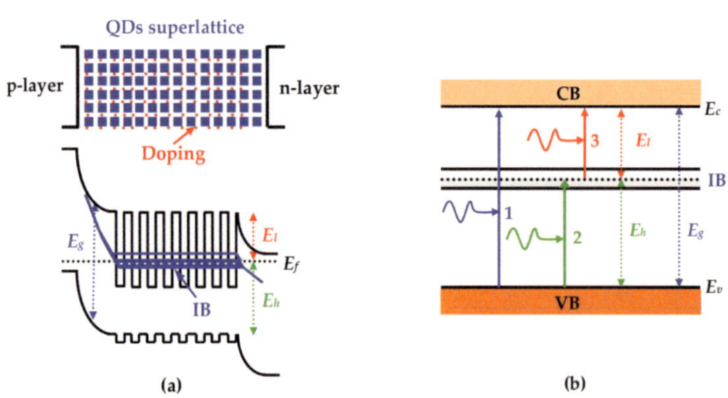

Figure 12. (a) Structure of QDIBSC and energy bands with possible photoabsorption and (b) simplified bandgap energies with three relevant optical transitions [45].

http://dx.doi.org/10.5772/67405

SCs [44]. In the last decade, there has been an extensive effort to demonstrate the QDIBSCs with a central focus on III-V compound semiconductors [45–53]. QDs must be dense to achieve the sufficient photo-absorption in QD layers. QDs are very small and do not absorb much light. Thus, it is necessary to have several layers of QDs. But, these layers cause additional strain and, therefore, the substrate of SC is damaged and performance is reduced. Therefore, the number of QD layers is limited. The typical real density of QDs is the order of 10^{10} per cm^2 in a single layer for InAs QDs grown on GaAs substrate [47]. For multi-stacked InAs/GaAs SCs with 10 and 50 stacked QDs layers, shown in **Figure 13**, the QDs are vertically aligned in the growth direction without dislocations in 10 stacked QD layers. By contrast, a lot of strain-induced dislocations are observed at both the GaAs substrate and InAs QDs grown regions on sample with 50 stacked QD layers. These defects lead to a significant non-radiative recombination and a reduction in photocurrent productions. Efficiencies of this configuration are now greater than 18%. By using the intermittent deposition technique, ultra-high stacked $In_{0.4}Ga_{0.6}As$/GaAs QDIBSC fabricated [47]. The critical thickness in $In_{0.4}Ga_{0.6}As$/GaAs system is much thicker than in InAs/GaAs system. Therefore, $In_{0.4}Ga_{0.6}As$ QDs have the advantage of being able to ultra-high stack. In this type, no dislocations are generated after the stacking up to 300 QD layers. They have been successful in the stacking of 400 QD layers while keeping the high crystal quality.

More theoretical calculations have shown that efficiencies of 63% for IBSCs can be achieved under maximum sun concentration and 47% for one sun, which is a significant improvement on the corresponding maximum single junction efficiency of 41% [44–47]. The reason for improvement of the efficiency in QDIBSCs is mainly that the IB formed among QDs increases absorption of longer wavelength region sunlight and reduces non-radiative combination. Recent studies have shown that efficiencies of QDIBSCs can be improved by using different compositions from III-V semiconductor materials [48–53]. It depends essentially on controlling each of the QDs size and the distances between them in the host materials. For example, in reference [48], the authors designed and studied a theoretical model for one intermediate band QDs SC. The composition of both QDs and host materials are $InAs_{0.9}N_{0.1}$ and $GaAs_{0.98}Sb_{0.02}$, respectively. These studies are based on the Schrödinger equation and it is solved by using the Kronig-Penney model. This work changes the size of QDs and the distance between them, and therefore controls of the location of the IB between the CB and VB. The authors have changed the size of QDs by the values of 8.1, 8.6 and 9.1 nm with the distance between the QDs in the host material constant at 1.98 nm.

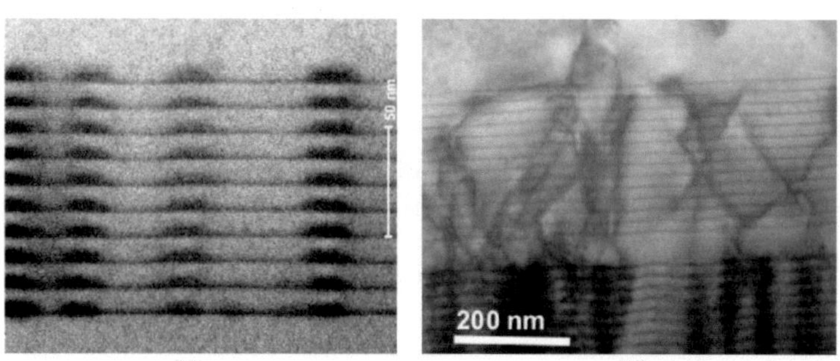

Figure 13. QDIBSC with 10 QD layers in (a) and 50 QD layers in (b) from InAs grown on GaAs substrate [47].

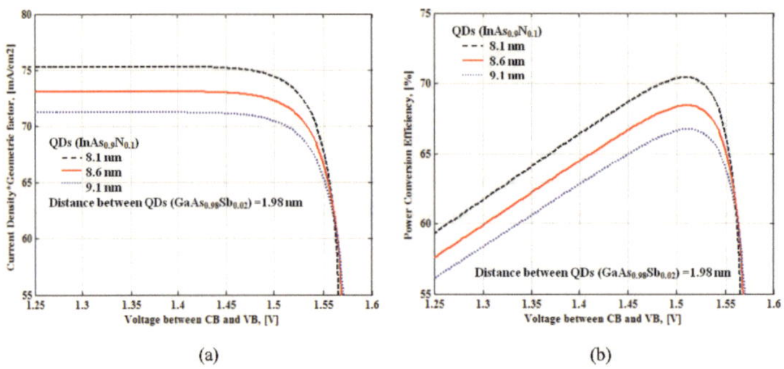

Figure 14. Current density in (a) and output power conversion efficiency in (b) for QDIBSC at maximum concentration [48].

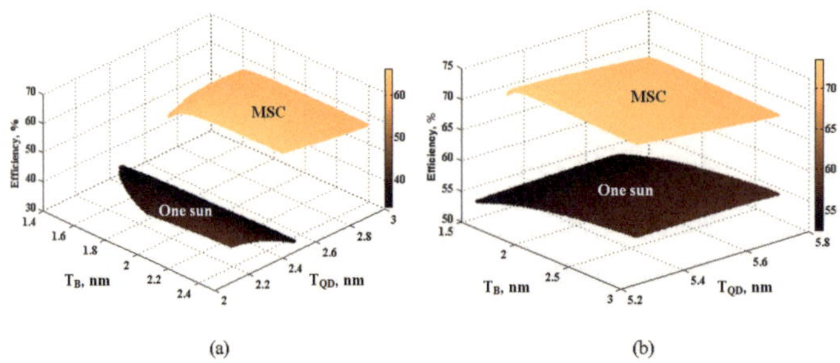

Figure 15. Efficiencies versus size of QDs (T_{QD}) and distance between QDs (T_B) for one IB and two IBs QDSC at one sun and maximum sun concentration (MSC) [53]. One-IB QDSC at one sun and MSC. Two-IBs QDSC at one sun and MSC.

They obtained a relationship between the cell voltage and the output current density from one side and the power conversion efficiency from another side, as shown in **Figure 14(a)** and**(b)**.

From **Figure 14**, it is clear that the current density and efficiency of the cell change inversely with change of the QDs size. The efficiency reached 57.5% and 70.4% at one sun and maximum concentration, respectively [48]. One of the challenges of the QDIBSCs theoretical study is to gather between the one and two IBs in the same time as it is a very complicated issue because some parameters should achieve the physical nanostructure layer concepts. One of the important parameter which should be preserved during theoretical derivation is potential bandgap energy condition. *Aly* and *Nasr* recently studied these cases (one and two IBs) to determine which one will yield highest efficiency [53]. From **Figure 15**, one can notice that the highest efficiency is achieved in the case of two IBs and reached 73.55% at maximum concentration with the following data: size of QDs is 5.25 nm, distance between them is 2.09 nm, and the composition of both QDs and host materials are $InAs_{0.9}N_{0.1}$ and $GaAs_{0.98}Sb_{0.02,}$ respectively.

5. Conclusions

In this chapter, current attempts for the use of nanostructure materials to improve the performance of SCs have been reviewed. Some of the different ways to reduce the cost and increase the efficiency of SCs have been summarized. Nanostructures for SCs that have been discussed in this work are nanocomposites, QWs, QRs and QDs. Nanocomposite SCs based on thin films, hot carrier and dye-sensitized structures have shown promising performance in commercial context. Nanostructured layers in thin film SCs offer a number of important advantages such as the effective optical path for absorption is much larger than the actual film thickness, recombination losses are greatly reduced, the absorber layer thickness can be as thin as 150 nm and thus installation costs can be reduced and finally, reduced manufacturing costs as a result of using a low temperature process. The performance of hot carrier SCs improves when the absorber layer is QDs superlattice. For dye-sensitized SCs, the wide-bandgap energy semiconductors such as TiO_2 and CNT fibres in the form of nanoparticles have been used to maximize the contact area between them and the dye. The presence of QW in the depletion layer of SCs produces electric field that leads to the collection of charge carriers photo-generated in the wells, leading to an enhanced current. The efficiency of this type is improved when the number of QWs increases. QRs SCs improve the performance and reduce the cost based on increased absorption due to diffuse light scattering in QR arrays, short collection lengths of minority carriers and flexibility of cell integration on a variety of low-cost carrier substrates. Finally, QDs have been shown to be useful in SC devices in various modes such as multi-layered QD arrays and QDIBSCs. The efficiency for multi-junction achieved is 41%. Nanostructures are used to realize different bandgap energies using only one material such as Si to reduce the cost. Therefore, all-Si multi-junction tandem SCs can be produced by using different size nanocrystals or QDs from Si. The QDIBSCs are one of the most promising candidates to improve and enhance the performance of SCs. More theoretical studies for this type were carried by using different compositions from III-V semiconductor materials. In fact, these studies showed that the performance of this type depends largely on the compositions of materials, size of QDs and the distance between them in the host materials. The theoretical efficiency in these SCs reached as high as 73.55%.

Author details

Abouelmaaty M. Aly

Address all correspondence to: abouelmaaty67@gmail.com

1 Electronic Research Institute, Cairo, Egypt

2 College of Computer, Qassim University, Buryadah, Kingdom of Saudi Arabia

References

[1] Tsakalakos L: Nanostructures for photovoltaics. Materials Science and Engineering R. 2008; 62: 175-189. DOI: 10.1016/j.mser.2008.06.002

[2] Shockley W, Queisser H. J: Detailed balance limit of efficiency of p-n junction solar cells. Journal of Applied Physics. 1961; 32(3): 510-519. DOI: 10.1063/1.1736034

[3] Singh J: Electronics and Optoelectronic Properties of Semiconductor Structures. 1st ed. Cambridge, UK: Cambridge University Press; 2003. 288 p. DOI: 10.2277/052182379X

[4] Nayfeh M O, et al.: Thin film silicon nanoparticle UV photodetector. IEEE Photonics Technology. 2004; 16(8): 1927-1929. DOI: 10.1109/LPT.2004.831271

[5] Aly M A: Improving the Power Conversion Efficiency of the Solar Cells. Qassim University Journal of Engineering and Computer Sciences. 2014; 7(2): 135-156. http://publications.qu.edu.sa/ojs/index.php/enginerring/article/view/1399

[6] Zhang G, et al.: Semiconductor nanostructure-based photovoltaic solar cells. Nanoscale. 2011; 3(6): 2430-2443. DOI: 10.1039/C1NR10152H

[7] Zhang Q, Gao G: Nanostructured photoelectrodes for dye-sensitized solar cells. Nanotoday. 2011; 6(1): 91-109. DOI: 10.1016/j.nantod.2010.12.007

[8] Zhu J, et al.: Optical absorption enhancement in amorphous silicon nanowire and nanocone arrays. Nano Lett. 2009; 9(1): 279-282. DOI: 10.1021/nl802886y

[9] Kibria T M, et al.: A review: Comparative studies on different generation solar cells technology. In: Proceedings of 5th International Conference on Environmental Aspects of Bangladesh (ICEAB); 5-6 September 2014; Dhaka. Bangladesh; 2014. pp. 51-53.

[10] Ojajarvi J: Tetrahedral chalcopyrite quantum dots in solar cell applications [thesis]. Department of Physics: University of Jyvaskyla; 2010.

[11] Javawardena G D K, et al.: Inorganics-in-organics: recent developments and outlook for 4G polymer solar cells. Nanoscale. 2013; 5: 8411-8427. DOI: 10.1039/C3NR02733C

[12] Kasap S, Capper P: Springer Handbook of Electronic and Photonic Materials. In: Fox M, Ispasoiu R, editors. Quantum Wells, Superlattices and Band-Gap Engineering. 2nd ed. Springer:NY, US; 2007. pp. 1021-1040. DOI: 10.1007/978-3-319-48933-9.ch42

[13] Morozhenko V: Infrared Radiation. In: Nasr A, editor. Infrared Radiation Photodetectors. InTech:Rijeka, Croatia; 2012. pp. 85-126. DOI: 10.5772/2031.ch5

[14] Soga T: Nanostructured Materials for Solar Energy Conversion. 1st ed. Elsevier B. V.: Amsterdam, Netherlands; 2006. 485 p. DOI: 10.1016/B978-044452844-5/50000-7

[15] Deng Q, et al.: Theoretical study on $In_xGa_{1-x}N/GaN$ quantum dots solar cell. Physica B. 2011; 406(1): 73-76. DOI: 10.1016/j.physb.2010.10.020

[16] Singha R, et al.: Nanostructured CdTe, CdS and TiO_2 for thin film solar cell applications. Solar Energy Materials & Solar Cells. 2004; 82(1): 315-330. DOI: 10.1016/j.solmat.2004.02.006

[17] Nabhani N, Emami M: Nanotechnology and its Applications in Solar Cells. In: 2nd International Conference on Mechanical and Industrial Engineering (ICMIE'2013); 8-9 March 2013; Kota Kinabalu. Malaysia; 2013. pp. 88-91.

[18] König D, et al.: Hot carrier solar cells: principles, materials and design. Physica E: Low-dimensional Systems and Nanostructures. 2010; 42(10): 2862-2866. DOI: 10.1016/j.physe.2009.12.032

[19] Zhang Y, et al.: Development of inorganic solar cells by nanotechnology. Nano-Micro Lett. Springer Berlin Heidelberg, Germany. 2012; 4(2): 124-134. DOI:10.1007/BF03353703

[20] Conibeer J G, et al.: Slowing of carrier cooling in hot carrier solar cells. Thin Solid Films. 2008; 516(20): 6948-6953. DOI: 10.1016/j.tsf.2007.12.102

[21] Aliberti P, et al.: Investigation of theoretical efficiency limit of hot carriers solar cells with a bulk indium nitride absorber. Journal of Applied Physics. 2010; 108(094507): 1-10. DOI:10.1063/1.3494047

[22] Grätzel M: Dye-sensitized solar cells. Journal of Photochemistry and Photobiology C: Photochemistry Reviews. 2003; 4(2): 145-153. DOI: 10.1016/S1389-5567(03)00026-1

[23] Wang P, et al.: A stable quasi-solid-state dye-sensitized solar cell with an amphiphilic ruthenium sensitizer and polymer gel electrolyte. Nature Materials. 2003; 2: 402-407. DOI:10.1038/nmat904

[24] Yang J H, et al.: Characteristics of the dye-sensitized solar cells using TiO_2 nanotubes treated with TiCl4. Materials. 2014; 7(5): 3522-3532. DOI:10.3390/ma7053522

[25] Alturaif H A, et al.: Use of carbon nanotubes (CNTs) with polymers in solar cells. Molecules. 2014; 19: 17329-17344. DOI:10.3390/molecules191117329

[26] Zhang Y, et al.: Improved dye sensitized solar cell performance in larger cell size by using TiO_2 nanotubes. Nanotechnology. 2013; 24(4): 045401-6. DOI:10.1088/0957-4484/24/4/045401

[27] Carlos I. Cabrera, et al.: Modeling multiple quantum well and superlattice solar cells. Natural Resources. 2013; 4: 235-245. DOI: 10.4236/nr.2013.43030

[28] Carlos I. Cabrera, et al.: Modelling of GaAsP/InGaAs/GaAs strain-balanced multiple-quantum well solar cells. Journal of Applied Physics. 2013; 113(024512): 1-7. DOI: 10.1063/1.4775404

[29] Yang M, Yamaguchi M: Properties of GaAs/InGaAs quantum well solar cells under low concentration operation. Solar Energy Materials and solar Cells. 2000; 60(1): 19-26. DOI: 10.1016/S0927-0248(99)00055-0

[30] Garnett E C, et al.: Nanowire solar cells. Annual Review of Materials Research. 2011; 41: 269-295. DOI: 10.1146/annurev-matsci-062910-100434

[31] Tsakalakos L, et al.: Silicon nanowire solar cells. Applied Physics Letters. 2007; 91: 233117-1 to 3. DOI: 10.1063/1.2821113

[32] Srivastava S K, et al.: Excellent anti-reflection properties of vertical silicon nanowire arrays. Solar Energy Materials and Solar Cells. 2010; 94(9): 1506-1511. DOI: 10.1016/j.solmat.2010.02.033

[33] Kim J Y, et al.: Postgrowth in situ chlorine passivation for suppressing surface-dominant transport in silicon nanowire devices. IEEE Transactions Nanotechnology. 2012; 11(4): 782-787. DOI: 10.1109/TNANO.2012.2197683

[34] Anna D M: Nanowire-based solar cells: Device design and implementation [thesis]. Ecole Polytechnique Federale de Lausanne, Suisse; 2014.

[35] Kerr M J, Cuevas A: General parameterization of auger recombination in crystalline silicon. Journal of Applied Physics. 2002; 91: 2473-2480. DOI: 10.1063/1.1432476

[36] Trupke T, et al.: Improving solar cell efficiencies by down-conversion of high-energy photons. Journal of Applied Physics. 2002; 92: 1668-1674. DOI: 10.1063/1.1492021

[37] Svrcek V, et al.: Silicon nanocrystals as light converter for solar cells. Thin Solid Films. 2004; 451(3): 384-388. DOI: 10.1016/j.tsf.2003.10.133

[38] Shalav A, et al.: Luminescent layers for enhanced silicon solar cell performance: up-conversion. Solar Energy Materials and Solar Cells. 2007; 91(9): 829-842. DOI: 10.1016/j.solmat.2007.02.007

[39] Marti A, et al.: Production of photocurrent due to intermediate-to-conduction-band transitions: a demonstration of a key operating principle of the intermediate-band solar cell. Physics Review Letters. 2006; 97(24): 247701-247704. DOI: 10.1103/PhysRevLett.97.247701

[40] Green M A, et al.: Nanostructured Silicon-Based Tandem Solar Cells [Project]. Sydney, Australia: University of New South Wales; 2005.

[41] Cho E-C, et al.: Silicon quantum dot/crystalline silicon solar cells. Nanotechnology. 2008; 19(24): 245201. DOI: 10.1088/0957-4484/19/24/245201

[42] Mavilla N R, et al.: Structural properties of ICPCVD fabricated SiO_2/SiOx superlattice for use in beyond Shockley-Queisser-limit solar cells. In: 27th European Photovoltaic Solar Energy Conference (EUPVSEC); 24-28 September 2012; Frankfurt. Germany; 2012. pp. 379-381.

[43] Rao M N, Solanki C S and Vasi J: Nanotechnology for next generation photovoltaics. In: 82nd Annual Symposium of NASI on "Nano-science and technology for Mankind"; 2012; Varanasi. India; 2012.pp. 340-350.

[44] Luque A, et al.: Increasing the efficiency of ideal solar cells by photon induced transitions at intermediate levels. Physics Review Letters. 1997; 78: 5014-5017. DOI: 10.1103/PhysRevLett.78.5014

[45] Nasr A, Aly A M. Performance evaluation of quantum-dot intermediate-band solar cell. Journal of Electronic Materials. 2016; 45(1): 672-681. DOI: 10.1007/s11664-015-4172-z

[46] Okada Y, et al.: Intermediate band solar cells: recent progress and future directions. Applied Physics Reviews. 2015; 2: 021302-1 to 48. DOI: 10.1063/1.4916561

[47] Martí A, et al.: Emitter degradation in quantum dot intermediate band solar cells. Applied Physics Letters. 2007; 90(233510): 6233-6237. DOI: 10.1063/1.2747195

[48] Aly A. M, Nasr A: Theoretical performance of solar cell based on minibands quantum dots. Journal of Applied Physics. 2014; 115(114311): 114311-1 to 113411-9. DOI: 10.1063/1.4868982

[49] Aly A, Nsar A: Theoretical study of one-intermediate band quantum dots solar cell. International Journal of Photoenergy:2014; 2014(904104): 1-10. DOI: 10.1155/2014/904104

[50] Aly A M: Progress into power conversion efficiency for solar cells based on nanostructured and realistic spectra. Journal of Renewable Sustainable Energy. 2014; 6(023118): 023118-1 to 023118-9. DOI: 10.1063/1.4873132

[51] Nasr A, Aly A: Theoretical investigation of some parameters into the behavior of quantum dot solar cells. Journal of Semiconductor. 2014; 35(12): 124001-8. DOI: 10.1088/1674-4926/35/12/124001

[52] Aly A M: Investigation of some parameters which affects into the efficiency of quantum dot intermediate band solar cell. International Journal of Renewable Energy Research. 2014; 4(4): 1085-1093. http://www.ijrer.org/ijrer/index.php/ijrer/article/view/1714

[53] Aly A, Nasr A: The effect of multi-intermediate bands on the behavior of an $InAs_{1-x}N_x$/ $GaAs_{1-y}Sb_y$ quantum dot solar cell. Journal of Semiconductor. 2015; 36(4): 042001-6. DOI: 10.1088/1674-4926/36/4/042001

Chapter 11

Nanotechnology in Herbicide Resistance

Evy Alice Abigail and Ramalingam Chidambaram

Additional information is available at the end of the chapter

http://dx.doi.org/10.5772/intechopen.68355

Abstract

Herbicide market in agriculture is a multi-billion dollar industry with sophisticated multi-impact issues, with increased weed resistance at the topmost. Nanoherbicides under development in the current decade could be a new strategy to address all the problems caused by the conventional non-nanoherbicides. With polymeric nanoparticles often used as nanocarriers for herbicide delivery, the current era has seen the rise of new nanoparticles-based delivery systems. As the potential use of nanostructured materials enables the use of herbicides effectively and rules out the emergence of weed-resistant population at an early stage, these very desirable nanotechnological practices in agriculture are reviewed here.

Keywords: herbicide, nanoherbicide, nanotechnology

1. Introduction

The genetically acquired capacity of the weed population to survive a herbicide exposure under normal usage conditions could be stated as herbicide resistance. The resistance brings the illustration of Darwin's 'survival of the fittest' principle. In a population of weeds exposed to herbicide, only a few individuals develop resistance, while the rest dies due to the herbicide action. This set resistant weed that survives eventually becomes a population of weeds with acquired resistance to a particular herbicide. The uncontrolled and repeated application of same herbicide will also select resistance plants. In some cases, multiple resistances can also appear due to sequential selection. Over the globe, nearly 249 herbicide-resistant weedy biotypes have been identified in over 47 countries. This number constantly grows on an annual basis giving rise to new resistant weeds. Likely, some management practices also give a rise to the development of herbicide-resistant weeds.

The herbicide-resistant management mostly involves a proper weed control program, with the aid of strategies such as usage of necessary herbicide specifically at recommended dosage rate, rotation of herbicides within herbicide groups and usage of herbicides mixtures. These strategies are followed for achieving the goal of effective control of resistance weed population. The system of integrated weed management (IWM) combines the application of all the possible weed control tools with accompanied economic crop production. Among the used tactics for weed resistance, the use of herbicide mixtures and rotations was found to be most useful. As weeds also have a hectic period to adapt to the management practice that keeps changing, the reliability on the cultural control could be of great importance. In spite of the reliance on the resistant management strategies, if the overuse of herbicides is controlled, then herbicide resistance will soon become archaic.

Nanotechnology offers exciting ways for averting the herbicide overuse and also a safe and effectual delivery [1]. The usage of nanostructured systems in agriculture has increased tremendously in the current era for the controlled release of agrochemicals as well for plant nutrients (**Figure 1**). The nanostructured herbicide could substantially reduce the herbicide consumption rate and promise increased crop productivity. This technology of exploiting nanomaterials guarantees to improve the current agricultural practices via the enrichment of management methods [2]. Nanoherbicides are one of the new-fangled strategies for combating the problems of conventional herbicides. These are being developed for addressing the issues in annual weed management and also for fatiguing the weed seed collection. The nanostructured formulation performs action through controlled release mechanism. The nanoherbicides comprise a wide range of entities such as polymeric and metallic nanoparticles. Nanoherbicides require a glance in order to place nanotechnology at the premier level.

Advancements in nanotechnology could be boon for mitigating the unsolvable herbicide resistance prevailing for centuries (**Figure 2**). The high penetration efficiency of nanoherbicides helps in eliminating the weeds before resistance could develop. The nanocarriers required for preparing nanoherbicides provide short- and long-residual herbicides based on the need by averting the lethal dose at which the plant could develop herbicide resistance. The preparation of nanoformulation with appropriate carriers would provide a basis for sustainable and economic agriculture. Nanoherbicides will start a high localization of the active substances only within the target plants avoiding the evolution resistance to particular herbicide at the basic level. Hence, the application of these nanotechnology-based miracle workers, nanoherbicides, for combating the herbicide resistance evolution is prodigious.

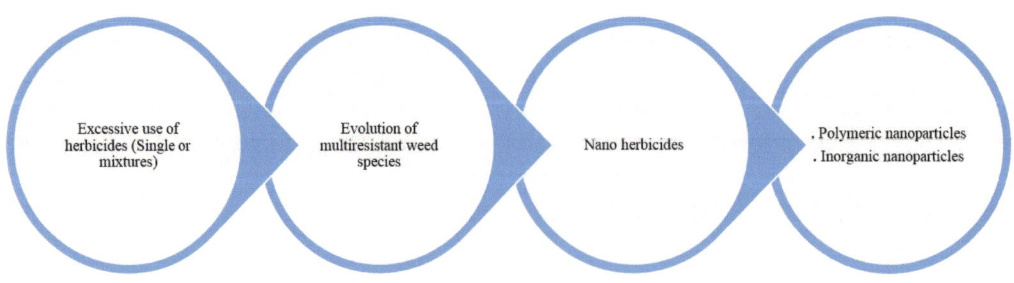

Figure 1. Hierarchy of the nanoherbicides evolution.

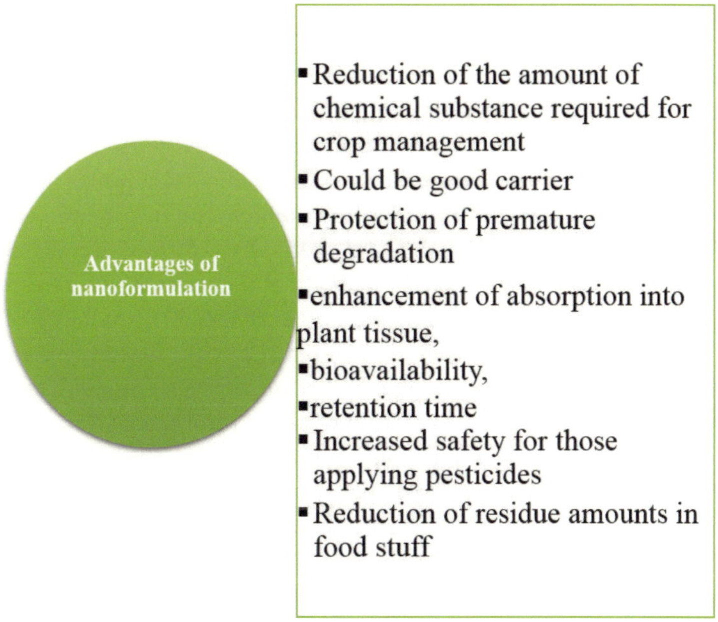

Figure 2. Advantages of nanoformulation over conventional herbicide formulation.

2. Nanoherbicides

Nanoherbicides are formulated by exploiting the nanotechnological potential for effectual delivery of chemical or biological pesticides with the help of nanosized preparations or nanomaterials-based herbicide formulations. Nanomaterials or nanostructures materials-based formulations could improve the efficacy of the herbicide, enhance the solubility and reduce the toxicity in comparison with the conventional herbicides. Early weed control with the use of nanoparticle-based herbicide release systems could reduce the herbicide resistance potential, maintain the activity of the active ingredient and prolong their release over a longer period [3]. The development of specific herbicide molecule encapsulated with nanoparticle aims at specific receptors present at the root of the targeted weed. The developed nanoparticle enters the root system of the weed and gets translocated to perform its action which in turn inhibits the glycolysis of the plant root system. The targeted action creates starvation of the plant and thus kills it. These nanoherbicides could also be used in rain-fed areas where herbicides get dissipated through vapourisation due to insufficient soil moisture. With the help of controlled release of herbicides via encapsulation, the weeds can be utterly destroyed. Apart from herbicides, adjuvants normally used to enhance the herbicidal activity are currently claiming to include nanomaterials. A glyphosate-resistant crop was reported to be made susceptible to glyphosate upon addition of a nanotechnology-based surfactant on to a soybean micelle. Nanoparticles can act as good carrier and also can form nanoformulation when added with herbicides. These nanoformulations assist in overcoming the main drawback of herbicide industry such as evolution of herbicide resistant plants. The nanoparticle systems for herbicide delivery are mostly composed of polymeric substances which are biodegradable with non-toxic metabolites.

2.1. Polymeric nanoparticles

Among the various types of nanoparticles used for formulating nanoherbicides, polymeric nanoparticle prepared either in the form of nanospheres or nanocapsules, is the most attractive form. Poly(epsilon caprolactone) is one polymer repeatedly used for the encapsulation of atrazine herbicide. Poly(epsilon caprolactone) possesses good physico-chemical properties along with enhanced bioavailability and biocompatibility. The polymeric nanoparticles containing atrazine herbicide were prepared and were characterized for size, polydispersity index, pH and encapsulation efficiency. The stability of the nanoparticles was found to be for a period of 3 months. The nanoparticle formulations reduced the mobility of the herbicide in soil but enhanced its herbicidal activity in comparison with free atrazine [4]. When tested against target plant, *Brassica* sp., the polymeric nanoparticles encapsulated with atrazine were proven effective. In another study, Grillo et al. [5] used the polymer for encapsulated three triazine herbicides such as atrazine, ametryn and simazine to reduce the environmental impact caused by them. The encapsulated polymeric nanoparticles of triazines possessed better association efficiency over 84%. The nanoparticles were found to have stability of size, zeta potential, pH and polydispersity for nearly 270 days. The triazine herbicide formulations revealed that the nanocapsules release the triazine *via* controlled release by relaxing the polymeric chains in vitro release experiments. The polymeric herbicide nanoparticles showed relatively less genotoxicity in *Allium cepa* chromosome aberration assay.

Alginate/Chitosan (Ag/Cs) nanoparticles were chosen for the encapsulation of parquet herbicide [6]. This polymeric complex has simple preparation methods adding further to their alternative use in agricultural applications. The Ag/Cs nanoparticle carrier system showed significant difference in the release profiles of free paraquat and the herbicide nanoparticles. The herbicide nanocarrier has altered the interaction of the herbicide in soil and indicated the effective means of averting the negative impacts of the herbicide induced by paraquat herbicide. Soil sorption studies with Ag/Cs herbicide paraquat nanoparticles exhibited dependence on the quantity of present soil organic matter. The enhanced interaction of paraquat released from Ag/Cs system in comparison with free paraquat revealed the effective of these polymeric nanoparticles as an excellent choice for eliminating the herbicide usage–associated ill effects. In a different study, the paraquat herbicide was encapsulated onto chitosan/tripolyphosphate nanoparticle and was proved efficient with this polymeric nanoparticle system as well. The herbicidal efficiency of paraquat was not found reduced even after encapsulated with very less toxicity. Cell culture viability tests and *Allium cepa* chromosomal aberration tests testified to the increased safety of the polymeric herbicide systems against non-target organisms [7]. Few works reported till date on polymeric nanoparticles encapsulated with herbicide provides a safe basis for using herbicides by reducing the adverse environmental impacts caused by them on human health as well on the environment.

2.2. Inorganic nanoparticles

Silica dioxide nanoparticles (SiNP) were explored as inorganic herbicide carriers in the recent past for active substances which are pH sensitive. These SiNPs maintain optimal herbicide concentrations with accompanied reduction in frequency of herbicide consumption rate.

These systems protect and stabilize the herbicide and reduce their wastage with easy deposit on the plant leaves. Rani et al. [8] stated the possible use of silica nanoparticles as herbicide carriers via a dynamic adsorption mechanism and sustained release. Though hypothesized, the use of SiNPS as herbicide carrier is yet to be explored in means of leaching behaviour, controlled release and toxicity.

2.2.1. Agro industrial waste based nanoparticles

In a study by Abigail et al. [9], an attempt was made to nanosized rice husk waste and was used as nanocarrier for 2,4-dichlorophenoxyacetic acid herbicide (2,4-D). The rice husk waste was brought down to nanosize and was surface adsorbed with 2,4-D to act as herbicide nanocarrier. The rice husk nanocarrier showed enhanced and reversible sorption of 2,4-D, illustrating its uniqueness to act as good carrier for encapsulating herbicides. The adsorption of 2,4-D on to the rice husk was not found to minimize the herbicidal activity when tested against target weed, *Brassica* sp. in comparison with free 2,4-D (**Figure 3**). The rice husk-based herbicide delivery system could be a boon due to its multiple advantages of utilizing the waste constructively apart from effectively using the herbicide with environmental contamination. Thus, the evolution of weeds with acquired resistance due to the uncontrolled use of herbicides could be eradicated with the help of these nanostructured herbicide carriers.

Figure 3. Bioactivity of rice husk-based 2,4-D nanoformulation against target weed *Brassica* sp. in comparison with other conventional methods [9].

3. Conclusions and future directions

Thus, in the current scenario, the overuse of herbicides to boost the crop production has left the soil, ground water and food products polluted. Although increasing the agricultural products is vital, the indirect damage on the environment cannot be unnoticed. Nanotechnology with promising results in the agricultural sector with its unique way of applying the pesticides, fertilizers etc., could enable the human population to finally visualize the dream of attaining sustainable and eco-friendly agricultural technology. This dream of exploiting the nanotechnological methods in agriculture is still in nascent stage. Therefore, development of systems that would improve the release profile of herbicides without altering their characteristics and novel carriers with enriched activity without significant environmental damage is the focus areas that require further investigations.

Author details

Evy Alice Abigail and Ramalingam Chidambaram*

*Address all correspondence to: nanobiolab115@gmail.com

Food Technology Lab, School of Biosciences and Technology, VIT University, Vellore, India

References

[1] Gonzalez JOW, Gutierrez MM, Ferrero AA, Band BF. Essential oils nanoformulations for stored-product pest control-characterization and biological properties. Chemosphere. 2014;**100**:130-138

[2] Sekhon BS. Nanotechnology in agri-food production:An overview. Nanotechnology, Science and Applications. 2014;**7**:31-53

[3] Manjunatha SB, Biradar DP,Aladakatti YR. Nanotechnology and its applications in agriculture: A review. Journal of Farm Sciences. 2016;**29(1)**:1-13

[4] Anderson ES. Pereira, Grillo R, Nathalie FS. Mello, Andre H. Rosa, Leonardo F. Fraceto. Application of poly(epsilon-caprolactone) nanoparticles containing atrazine herbicide as an alternative technique to control weeds and reduce damage to the environment. Journal of Hazardous Materials. 2014;**268:**207-215

[5] Grillo R, dos Santos NZP, Maruyama CR, Rosa AH, de Lima R, Fraceto LF. Poly(-caprolactone)nanocapsules as carrier systems for herbicides: Physico-chemical characterization and genotoxicity evaluation. Journal of Hazardous Materials. 2012;**231-232**:1-9

[6] dos Santos Silva M, Cocenza DS, Grillo R, de Melo NFS, Tonello PS, de Oliveira LC, Cassimiro DL, Rosa AH, Fraceto LF. Paraquat-loaded alginate/chitosan nanoparticles: Preparation, characterization and soil sorption studies. Journal of Hazardous Materials. 2011;**190**:366-374

[7] Grillo R, Pereira AE, Nishisaka CS, de Lima R, Oehlke K, Greiner R, Fraceto LF. Chitosan/tripolyphosphate nanoparticles loaded with paraquat herbicide: An environmentally safer alternative for weed control. Journal of Hazardous Materials. 2014;**278**:163-171

[8] Rani PU, Madhusudhanamurthy J, Sreedhar B. Dynamic adsorption of a-pinene and linalool on silica nanoparticles for enhanced antifeedant activity against agricultural pests. Journal of Pest Sciences. 2014;**87**:191-200

[9] Abigail MEA, Melvin Samuel S, Chidambaram R. Application of rice husk nano-sorbents containing 2,4-dichlorophenoxyacetic acid herbicide to control weeds and reduce leaching from soil. Journal of the Taiwan Institute of Chemical Engineers. 2016;**63**:318-326